PROGRESS IN TOTAL SYNTHESIS

Volume 1

PROGRESS IN

PROGRESS IN TOTAL SYNTHESIS

is one of a series of advanced titles published in cooperation with Landsberger Publishing Corporation.

TOTAL SYNTHESIS

Volume 1

Sarah Etheredge Danishefsky
Samuel Danishefsky

Department of Chemistry
University of Pittsburgh
Pittsburgh, Pennsylvania

APPLETON-CENTURY-CROFTS
Educational Division
MEREDITH CORPORATION
New York

ISBN 978-1-4684-8186-0   ISBN 978-1-4684-8184-6 (eBook)
DOI 10.1007/978-1-4684-8184-6

Copyright © 1971 by **MEREDITH CORPORATION**

Softcover reprint of the hardcover 1st edition 1971

731-1

Library of Congress Catalog Number: 72-150496

390-25325-1

To our children
Susannah and Daniel

ACKNOWLEDGMENTS

We express our genuine appreciation to past and present members of the group who have joined in the project. These are Thomas Bryson, Gary Koppel, Bruce Migdalof, George Roynyak, and Lantz Crawley, who have, in the interim, completed their studies at the university, and James Eggler, Ellis Hatch, and Arthur Nagel, who are still in residence.

We also wish to thank Dr. M.P. Etheredge, Professor Emeritus, Mississippi State University, for the author index, Mrs. Karen Nagel and Mrs. Reah Wender for their patient and skillful typing, and Miss Barbara Peterson for her proofreading services.

PREFACE

This series stemmed from a group of weekly seminars in our research group aimed at keeping its members abreast of recent developments in organic synthesis. The seminars tended to consist of several syntheses of natural products or related systems with particular emphasis on the general strategy inherent in the effort, new and interesting reactions which were utilized in the work, and specificity (or the lack of it) in arranging the relative stereochemistry of asymmetric centers and the geometry of double bonds.

We found that natural products offered an attractive setting in which the larger science of organic chemistry could be put to crucial tests. A truly elegant synthesis is a major advance in that it epitomizes how an imaginative mastery of the course of organic reactions can achieve a sophisticated objective by an economy of operations. Indeed any successful synthesis of a reasonably complex product, however cumbersome and graceless, is an important event for those who delight in the problem-solving dimension of science.

It is our hope and expectation that this yearly series of books will be of service at various levels. For those harried synthesis practitioners who experience difficulty in metabolizing the avalanche of literature into their research or teaching programs, we have attempted to maximize the information and stimulation per unit page by setting forth the essence of the synthesis in intelligible form. Such readers can gain a quick visual grasp of the salient issues. References are provided on the pages to lead such readers back to the original papers for detailed information on yields and reaction conditions. Surely such delving will be necessary if the information perceived through our illustrations is to be put to direct use.

For those organic chemists whose interest in synthesis is not sufficient to justify reading the original literature, these volumes could provide a quick (and we hope pleasant) method of keeping in touch with the general trends in the area.

For those who are beginning their study of organic synthesis, we believe that realistic exposure to the actual methods by which organic chemists solve synthetic problems is a more effective learning device than the antiseptic, classical approach of Name Reactions, however artfully organized. In the course of studying these syntheses, the beginner will be exposed to an array of reactions encountered in a realistic context and executed by real-life chemists. We have utilized total synthesis as a teaching device for a combined upper level undergraduate and first year graduate course in organic synthesis at the University of Pittsburgh with great reward. To enhance the value of the book in this regard, we have included comments on each synthesis (with references and diagrams where we thought appropriate) designed to place it in a broader context and clarify its more puzzling features. Extensive cross-referencing is utilized to emphasize recurring synthetic themes.

The book is divided into three chapters. The first deals with the synthesis of alkaloids and related structures. The organization is alphabetical. The second chapter deals with terpenes. These are organized into the familiar subgroups of monoterpenes, sesquiterpenes, diterpenes and triterpenes. The presentation within each level is based first on the number of carbocyclic rings and then on alphabetical names. The third chapter deals with an assortment of natural products which are grouped (somewhat arbitrarily) into antibiotics and metabolites, cannibinols, juvenile hormones, porphyrins, prostaglandins, steroids, and zearalenones.

We have omitted highly important but specialized areas such as nucleotides, peptides, and carbohydrates (except for kasugomycin). The emphasis is on construction of rings and control over asymmetric centers rather than on specific coupling techniques.

Optically active products are indicated in the usual fashion. All other products which contain asymmetry are, of course, d,l mixtures; however, the (±) designation has been omitted, since it is applicable practically everywhere.

All of the primary papers appeared in the chemical literature in the year 1968. In several cases, papers which appeared in 1969 are cited in the comments as being particularly germane to the work. The 1969 literature will be surveyed in detail in Volume 2.

CONTENTS

Acknowledgments   vii

Preface   ix

I. Alkaloids and Intermediates in the Synthesis of Alkaloids

   1. ACRONYCINE   2
   2. AJACONINE AND ATIDINE INTERMEDIATES   4
   3. ANDROCYMBINE-TYPE COMPOUNDS   6
   4. APOFERROROSAMINE AND MYOSMINE   8
   5. 6,7-BENZOMORPHAN   10
   6. CONDYFOLINE, TUBIFOLINE, AND TUBIFOLIDINE   12
   7. COREXIMINE   14
   8. (–)-CORYNANTHEIDINE   16
   9. (*3S,15S,20R*)-CORYNANTHEINE   18
  10. DASYCARPIDONE AND EPIDASYCARPIDONE   20
  11. DASYCARPIDONE AND EPIDASYCARPIDONE   22
  12. DIHYDROCRININE AND DIHYDROVITTATINE   24
  13. DIHYDROGAMBIRTANNINE AND ASPIDOSPERMA-STRYCHNOS
     ALKALOID MODELS   26
  14. 15,16-DIMETHOXYERYTHRINAN-3-ONE   28
  15. SYNTHETIC APPROACHES TO DITERPENE ALKALOIDS   30
  16. DITERPENE ALKALOID INTERMEDIATES   32
  17. DITERPENE ALKALOIDS   34
  18. EPIIBOGAMINE AND IBOGAMINE   36
  19. EPIIBOGAMINE   38
  20. SYNTHESIS OF RING SYSTEMS RELATED TO HASUBANAN ALKALOIDS   40
  21. HERNADINE   42
  22. HOMOERYTHRINADIENONE   44
  23. HOMOPROAPORPHINE-TYPE COMPOUNDS   46
  24. β,β-DISUBSTITUTED INDOLINE DERIVATIVES   48
  25. 17-ACETYL-5 α-ETIOJERVA-12,14,16-TRIEN-3β-OL   50
  26. LUPININE AND EPILUPININE AND THE HYDROLULOLIDINE AND
     HYDROJULOLIDINE RING SYSTEMS   52
  27. 12-EPILYCOPODINE   54
  28. LYCOPODINE   55
  29. LYCOPODINE   56

30.	MESEMBRINE	59
31.	MESEMBRINE	60
32.	MESEMBRINE	61
33.	MESEMBRINE	62
34.	MESEMBRINE	63
35.	OCHOTENSINE AND OCHOTENSIMINE	65
36.	OCHOTENSIMINE	66
37.	PERLOLIDINE	68
38.	PETALINE	70
39.	QUEBRACHAMINE SKELETON	72
40.	SANGUINARINE	74
41.	SEMPERVIRINE HYDROBROMIDE	76
42.	SOLANIDINE	78
43.	STEBISIMINE	80
44.	VEATCHINE	82
45.	VELBANAMINE	84
46.	VERARINE	86
47.	MONOMERIC VINCA ALKALOIDS—VINCADINE, VINCAMINOREINE, VINCAMINORINE, VINCADIFFORMINE, AND MINOVINE	88
48.	DIMERIC VINCA ALKALOIDS	90
49.	VINCAMINOL	92
50.	1-METHYL-16-DEMETHOXYCARBONYL-20-DESETHYLIDENE-VOBASINE	94
51.	WITHASOMNINE	96
52.	WITHASOMNINE	97

II. Terpenes and Related Structures
    A. MONOTERPENES

1.	ACTINIDIOLIDE, DIHYDROACTINIDIOLIDE, AND ACTINIDOL	100
2.	NEZUKONE	102
3.	ROSEFURAN AND DEHYDROELSHOLTZIONE	104
4.	VERBENALOL	106
5.	SABINENE, SABINA KETONE, AND SABINENE HYDRATE	108

    B. SESQUITERPENES

1.	ABSCISIC ACID	110
2.	JUVABIONE	112
3.	ANDROGRAPHOLIDE LACTONE	114
4.	AR-HIMACHALENE	116
5.	BULNESOL	118
6.	CARABRONE	120
7.	EREMOPHILENE AND EREMOLIGENOL	122
8.	EREMOPHIL-3,11-DIENE	124
9.	TETRAHYDROEREMOPHILONE	126
10.	β-EUDESMOL	128
11.	DEHYDROFUROPELARGONES	130
12.	ISOIRESIN, DIHYDROIRESIN, AND ISODIHYDROIRESIN	132
13.	JUNIPER CAMPHOR AND SELIN-11-EN-4α-OL	134
14.	NOOTKATONE	136
15.	VALERANONE	138
16.	β-VETIVONE	140
17.	NOR-KETOAGAROFURAN	142
18.	α-AGAROFURAN AND β-AGAROFURAN	144
19.	$\Delta'^{(10)}$-ARISTOLENE	146
20.	ARISTOLONE	148
21.	4-DEMETHYLARISTOLONE	150
22.	α-BOURBONENE AND β-BOURBONENE	152
23.	α-BOURBONENE	153
24.	A FUNCTIONALIZED ILLUDANE	154
25.	ILLUDIN M	155
26.	LINDESTRENE	156
27.	PATCHOULI AND EPIPATCHOULI ALCOHOLS	158
28.	SATIVENE	160

C. DITERPENES
    1. METHYL DEISOPROPYLDEHYDROBIETATE   162
    2. CARNOSOL DIMETHYL ETHER AND CARNOSIC ACID DIMETHYL ETHER   164
    3. FERRUGINOL   166
    4. FICHTELITE   168
    5. KAUR-16-EN-19-OIC ACID   170
    6. METHYL-2-HYDROXY-1,1,4a-TRIMETHYL-7-OXOPERHYDROPHENAN-
        THRENE-8a-CARBOXYLATE   172
    7. 10-CYANO-12-HYDROXY-7-OXO-17-NORPODOCARPA-5,8,11,13-
        TETRAENE   174
    8. DESOXYPODOCARPIC ACID AND 13-METHOXYPODOCARPIC ACID   176
    9. PODOCARPIC ACID   178
   10. SANDARACOPIMARIC ACID   180
   11. METHYL TRACHYLOBANATE   182
D. TRITERPENES
    1. β-AMYRINE   184
    2. CYCLOARTENOL   186

III. Other Natural Products
A. ANTIBIOTICS AND METABOLITES
    1. ANTHRAMYCIN   190
    2. CYCLOPENIN   192
    3. 6-DEMETHYL-6-DEOXYTETRACYCLINE   194
    4. TERRAMYCIN   196
    5. DIHYDRORADICININ   198
    6. FULVOPLUMIERIN   200
    7. GEOSMIN   202
    8. INDOLMYCIN   204
    9. KASUGAMYCIN   206
   10. 5,6-*Trans*-PENICILLIN V METHYL ESTER   208
   11. PYRETHROLONE   210
   12. RHODOQUINONE   212
B. CANNABINOLS
    1. ALLOEVODIONOL, CANNABICHROMENE, AND FRANKLINONE   214
    2. CANNABICHROMENE AND CANNABICYCLOL   216
    3. Δ6a,10a-TETRAHYDROCANNABINOL   218
    4. TETRAHYDROCANNABINOL CONGENER   220
C. JUVENILE HORMONES AND PROPYLURE
    1. STEREOSPECIFIC APPROACH TO JUVENILE HORMONES   222
    2. CECROPIA JUVENILE HORMONE   223
    3. CECROPIA JUVENILE HORMONE   224
    4. PROPYLURE   226
D. PORPHYRINS
    1. DEOXOPHYLLOERYTHROETIOPORPHYRIN   228
    2. OCTAETHYLPORPHYRIN   230
    3. SYNTHESIS OF OXYPORPHYRINS   232
E. PROSTAGLANDINS
    1. PROSTAGLANDINS E_1, A_1, $F_{1\alpha}$, AND $F_{1\beta}$ 234
    2. PROSTAGLANDINS OF THE E_1 AND F_1 SERIES   236
    3. PROSTAGLANDINS   238
    4. 7-OXAPROSTAGLANDIN $F_{1\alpha}$ 240
F. STEROIDS AND RELATED SYSTEMS
    1. (−)-17β-HYDROXY-$\Delta^{9\,(10)}$DES A-ANDROSTEN-5-ONE 242
    2. 3-METHOXY-17β-CARBOXYOESTRA-1,3,5(10),6,8-PENTAENE   244
    3. RETROPROGESTERONE AND DEHYDRORETROPROGESTERONE   246
    4. ESTRONE   248
    5. 16,17-DEHYDROPROGESTERONE   250
G. ZEARALENONES
    1. DIDEOXYZEARALANE   252
    2. ZEARALENONE   254
    3. ZEARALENONE   256
Index   259

CHAPTER ONE

Alkaloids and Intermediates in the Synthesis of Alkaloids

I-1 ACRONYCINE

J. R. Beck, R. Kwok, R. N. Booher, A. C. Brown, L. E. Patterson, P. Pranc, B. Rockey, and A. Pohland; *J. Am. Chem. Soc.*, **90**, 4706 (1968).

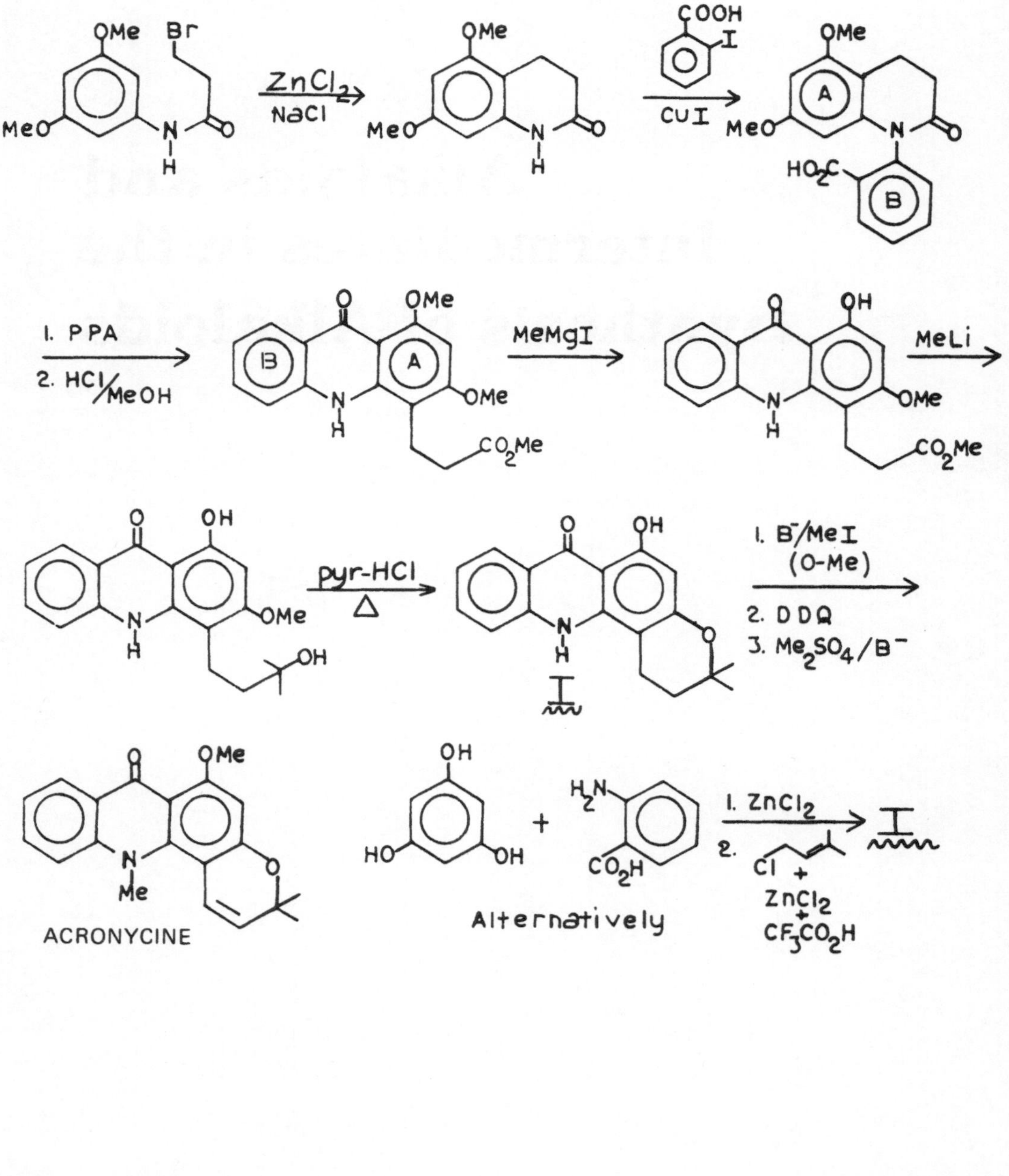

ACRONYCINE

Alternatively

I-1   The interesting feature of this synthesis is the formation of a lactam ring at a stage where the symmetry is such as to allow for only one product. When the lactam is opened, a required propionic acid residue is introduced at the α position of a dimethoxyacridone system. While the subsequent alternative synthesis obviated the need for this lengthy procedure, the strategy of using a ring closure-ring opening sequence to establish a position for a functional group is worthy of note. One of the more novel steps involves the use of a methyl Grignard reagent to cleave a phenolic methyl ester [cf. R. Mechoulam and R. Gaoni., *J. Am. Chem. Soc.*, 87, 3723, 1965]. Chelation of the magnesium with the proximate carbonyl group may be an important factor in differentiating between the two esters.

I-2 AJACONINE AND ATIDINE INTERMEDIATES

L. H. Zalkow, B. Kumar, D. H. Miles, J. Nabors, and N. Schnautz; *Tetrahedron Letters*, 1965 (1968).

I-2   The conversion of the olefinic epoxyacetate to the homoannular dienone may be formulated as follows:

The process is thus the sum of an epoxide-ketone rearrangement and a vinylogous β elimination.

The use of a succinic acid system as a precursor of an ethylene via oxidative bis decarboxylation with lead tetraacetate is a familiar technique whose most outstanding application came in the last step of the synthesis of Dewar benzene [see E. E. van Tamelen and S.P. Pappas, *J. Am. Chem. Soc.*, 85, 3297 (1963)]. An electrolytic variant has recently been shown to be advantageous [see P. Radlick, R. Klem, S. Spurlock, J. Sims, E. E. van Tamelen, and T. Whitesides, *Tetrahedron Letters*, 5117 (1968)].

I-3   ANDROCYMBINE-TYPE COMPOUNDS

T. Kametani, K. Fukumoto, F. Satoh, and H. Yagi, *J. Chem. Soc.*, (*C*) 3084 (1968).

ANDROCYMBINE

I-3   The key step in this synthesis is the thermal Pschorr-type ring closure which occurs with concurrent demethylation. Although the yield is low (1.5%), none of the homoapomorphine-type system was isolated. If none was, in fact, produced, the result is in contrast with that obtained from a lower homolog, using copper initiation (see Synthesis I-21). For a recent paper on the intricacies of the Pschorr reaction see A. H. Lewin and T. Cohen, *J. Org. Chem.*, 32, 3844 (1967).

I-4 APOFERROROSAMINE AND MYOSMINE

R. V. Stevens, M. C. Ellis, and M. P. Wentland; *J. Am. Chem. Soc.,* **90**, 5576 (1968).

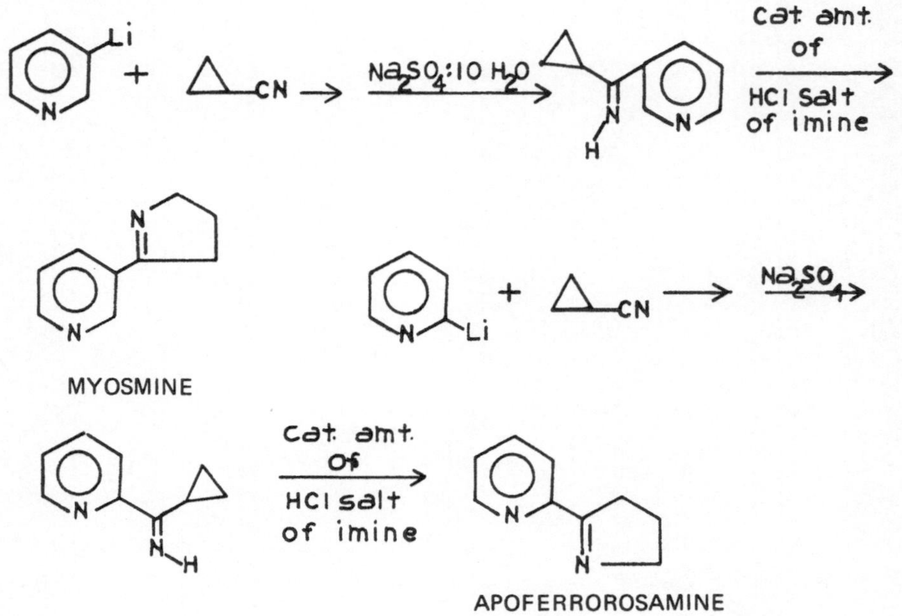

MYOSMINE

APOFERROROSAMINE

I-4   A cyclopropyl imine $\rightarrow \Delta^1$-pyrroline rearrangement is the crucial feature of these syntheses. When the nitrogen is substituted, a Δ^2-pyrroline is formed. The latter reaction is the basis of two syntheses of mesembrine (see I-30 and I-31).

I-5 6,7-BENZOMORPHAN

K. Kanematsu, R. T. Parfitt, A. E. Jacobson, J. H. Ager, and E. L. May; *J. Am. Chem. Soc.*, **90** 1064 (1968).

6,7-BENZOMORPHAN

I-5   This synthesis illustrates the utility of N-alkoxypyridinium salts for the purpose of introducing functionality at the 2 position of a pyridine nucleus.

Having been introduced, the cyano group is to be maintained while the pyridine ring is reduced to a piperidine. Quaternization of the nitrogen increases the vulnerability of the ring to reduction, thereby allowing for this objective. Having served its purpose, the extraneous N-methyl is removed by Von Braun degradation. An alternative demethylation procedure involved reaction of the tertiary amine with ethyl azodicarboxylate, followed by cleavage of the resultant product with pyridine hydrochloride.

I-6 CONDYFOLINE, TUBIFOLINE, AND TUBIFOLIDINE

B. A. Dadson, J. Harley-Mason, and (in part) G. H. Foster; *Chem. Communs.*, 20, 1233 (1968).

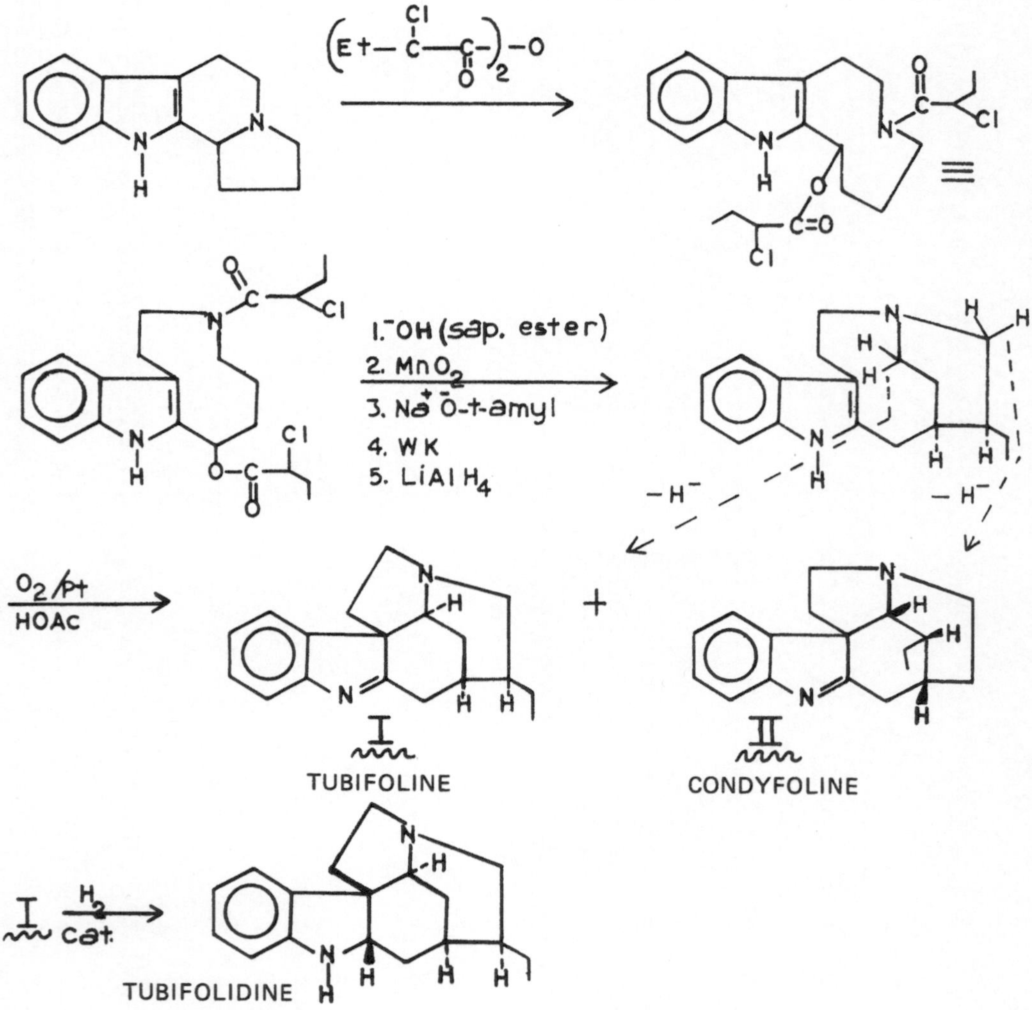

TUBIFOLINE

CONDYFOLINE

TUBIFOLIDINE

I-6   Electronic participation from the indolic nitrogen could well be a significant factor in the cleavage of the tertiary amine.

Whatever the mechanism, this reaction places the required α-chlorobutyryl group at nitrogen where it is readied for internal alkylation by a proximate enolate [cf. G. Stork and J. E. Dolfini, *J. Am. Chem. Soc.*, 85, 2872 (1963)]. The general notion of cleavage of a tertiary amine by acylation with an α-haloacetyl chloride followed by displacement with the corresponding α-halocarboxylate was employed to good advantage in the syntheses of dasycarpidone (see I-10).

Oxidation of the piperidine occurs in the two endocyclic possibilities to give, after Mannich closure at the β position of the indole, tubifoline and condyfoline (cf. I-47 where only one mode is reported).

I-7　COREXIMINE

T. Kametani, K. Fukumoto, H. Agui, H. Yagi, K. Kigasawa, H. Sugahara, M. Hiiragi, T. Hayasaka, and H. Ishimaru; *J. Chem. Soc.*, (*C*) 112 (1963).

COREXIMINE

I -7   As a result of a study of several benzyl isoquinoline substrates involving methoxy-hydroxyl permutations, the authors conclude that a free hydroxyl, in the *para* position to the point of cyclization, is required for this type of physiologically simulated (pH 7) berberine closure.

I-8 (-)-CORYNANTHEIDINE

C. S. Szántay and M. Bárczai-Beke; *Tetrahedron Letters*, 1405 (1968).

(-)-CORYNANTHEIDINE

I-8   The preferred conformation of the ethyl group in the alkylidene cyanoacetate appears to be axial so as to minimize repulsions with the *cis* β substituent on the double bond [cf. F. Johnson, *Chem. Rev.*, 68, 375 (1968)] . This factor is apparently responsible for changing the C-3—H:C-20—H relationship from *trans* to *cis* thus allowing for an equatorial disposition of the C-3 indole bond. The authors assume this to involve epimerization at C-3 but epimerization at C-20 is equally likely, *a priori.* Conjugate reduction occurs with an axial entry of hydride giving the equatorial dimethylmalonate residue at C-15.

I-9 *(3S, 15S, 20R)*-CORYNANTHEINE

R. L. Autrey and P. W. Scullard; *J. Am. Chem. Soc.,* **90,** 4917 (1968).

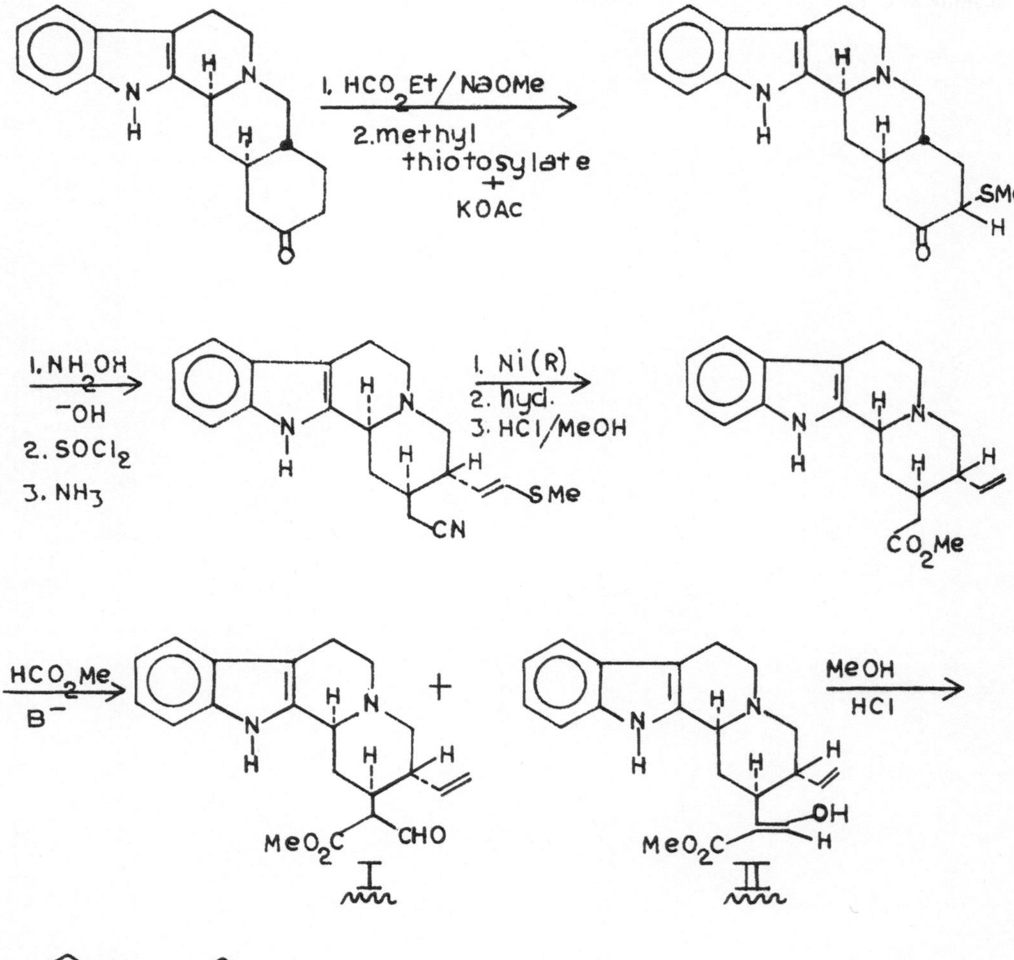

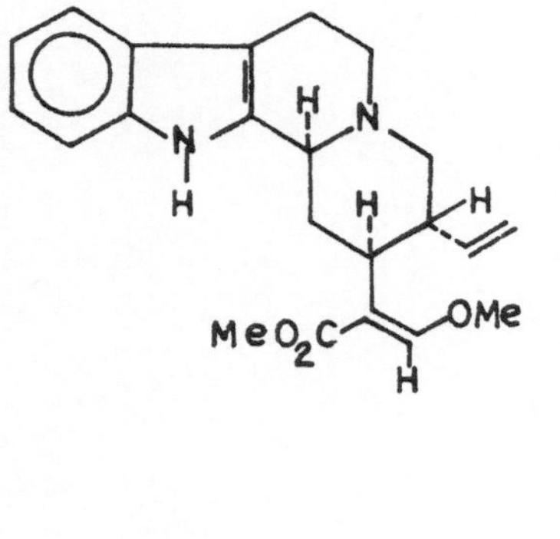

1. HCO₂Et / NaOMe
2. methyl thiotosylate + KOAC

1. NH₂OH / ⁻OH
2. SOCl₂
3. NH₃

1. Ni(R)
2. hyd.
3. HCl/MeOH

HCO₂Me / B⁻

MeOH / HCl

(3S, 15S, 20R)-CORYNANTHEINE

I-9   The synthesis of corynantheine from the yohimbone precursor illustrates the use of several synthetic stratagems. It will be noted that the *trans* fused D-E system serves as a means of insuring a *trans* relationship between the vinyl and β-methoxy-α-acrylyl functions. The *trans* junction also serves to orient the formylation to the desired side of the ketone [cf. L. Velluz, J. Valls, and G. Nominé, *Angew. Chem. Int. Ed. Engl.*, 4, 181 (1965)].

Cleavage is realized through a clever sulfur analog of a second order Beckmann reaction (cf. I-15 and I-16). The use of a thioenol ether as a progenitor of a vinyl group is another interesting feature of the synthesis.

I-10 DASYCARPIDONE AND EPIDASYCARPIDONE

L. J. Dolby and H. Biere; *J. Am. Chem. Soc.*, 90, 2699 (1968).

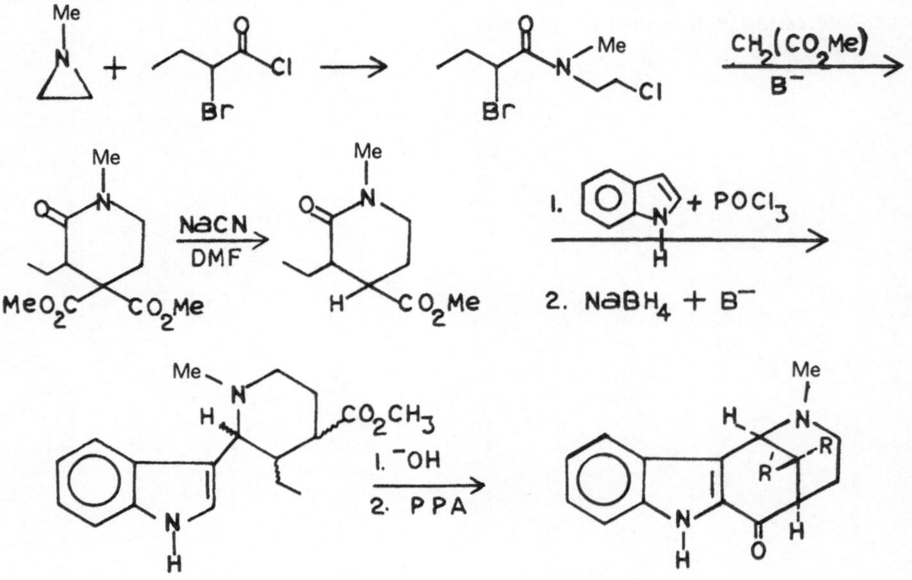

Mixture

R′ = Et; R = H—Dasycarpidone

R′ = H; R = Et—Epidasycarpidone

I-10   The acylation opening (cf. I-6) of a tertiary nitrogen is a key feature of the Dolby synthesis of dasycarpidone and epidasycarpidone. This produces a precursor for the desired 2-piperidone system. The latter is attached to indole by a Vilsmeier condensation followed by reduction. Decarbomethoxylation of the disubstituted malonic ester through the agency of cyanide is another noteworthy feature of the synthesis. Polyphosphoric acid closure of a carboxylic acid to the α-indole position is also used in Syntheses I-39 and I-50.

I-11 DASYCARPIDONE AND EPIDASYCARPIDONE

A. Jackson, A. J. Gaskell, N. D. Wilson, and J. A. Joule; *Chem. Communs.*, 364 (1968).

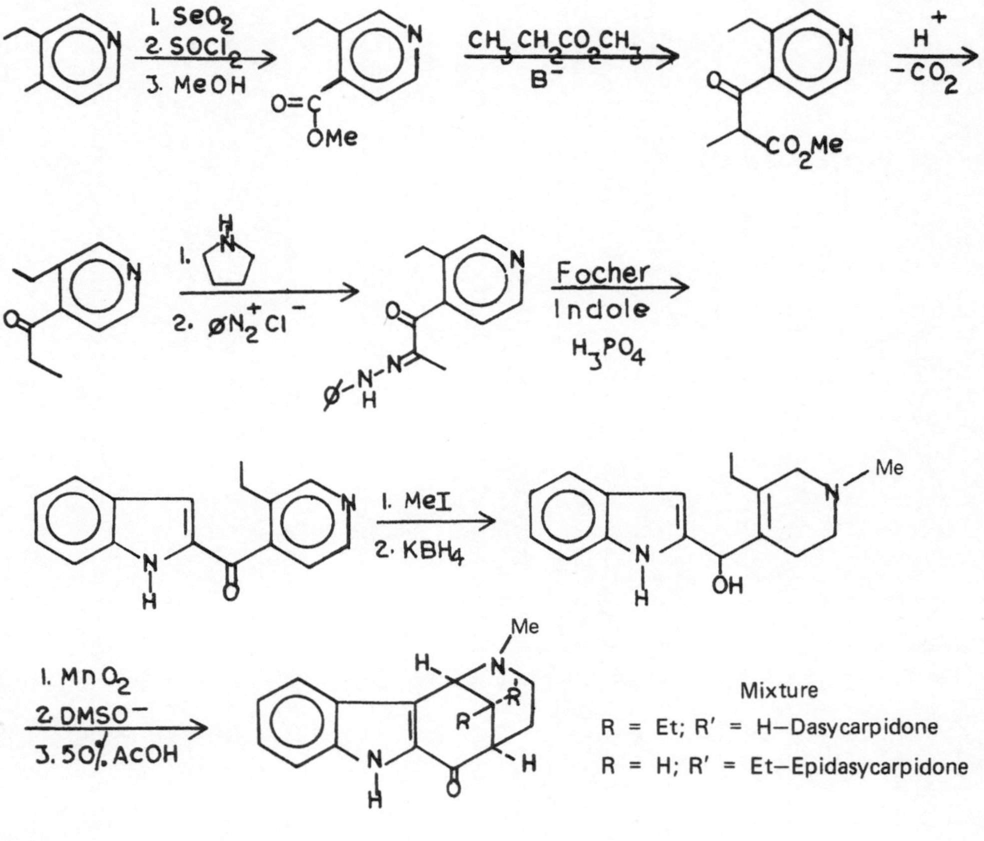

Mixture

R = Et; R′ = H—Dasycarpidone
R = H; R′ = Et—Epidasycarpidone

Mixture

I-11   This synthesis of dasycarpidone is totally different in design from that employed in I-10. Here a 3-ethyl-4-propionylpyridine system is used as a precursor of the piperidine system. The reaction of the enamine with benzenediazonium chloride constitutes an elegant method of preparing a specific monophenylhydrazone of an α-diketone. Note should also be taken of the utilization of the α carbon of an enamine as the electrophile in attacking the β position of the indole. This electrophilicity is no doubt generated by reversible β protonation of the enamine which sets the stage for a Mannich closure (cf. I-6).

I-12 DIHYDROCRININE AND DIHYDROVITTATINE

H. Irie, S. Uyeo, and A. Yoshitake; *J. Chem. Soc.*, (*C*), 1802 (1968).

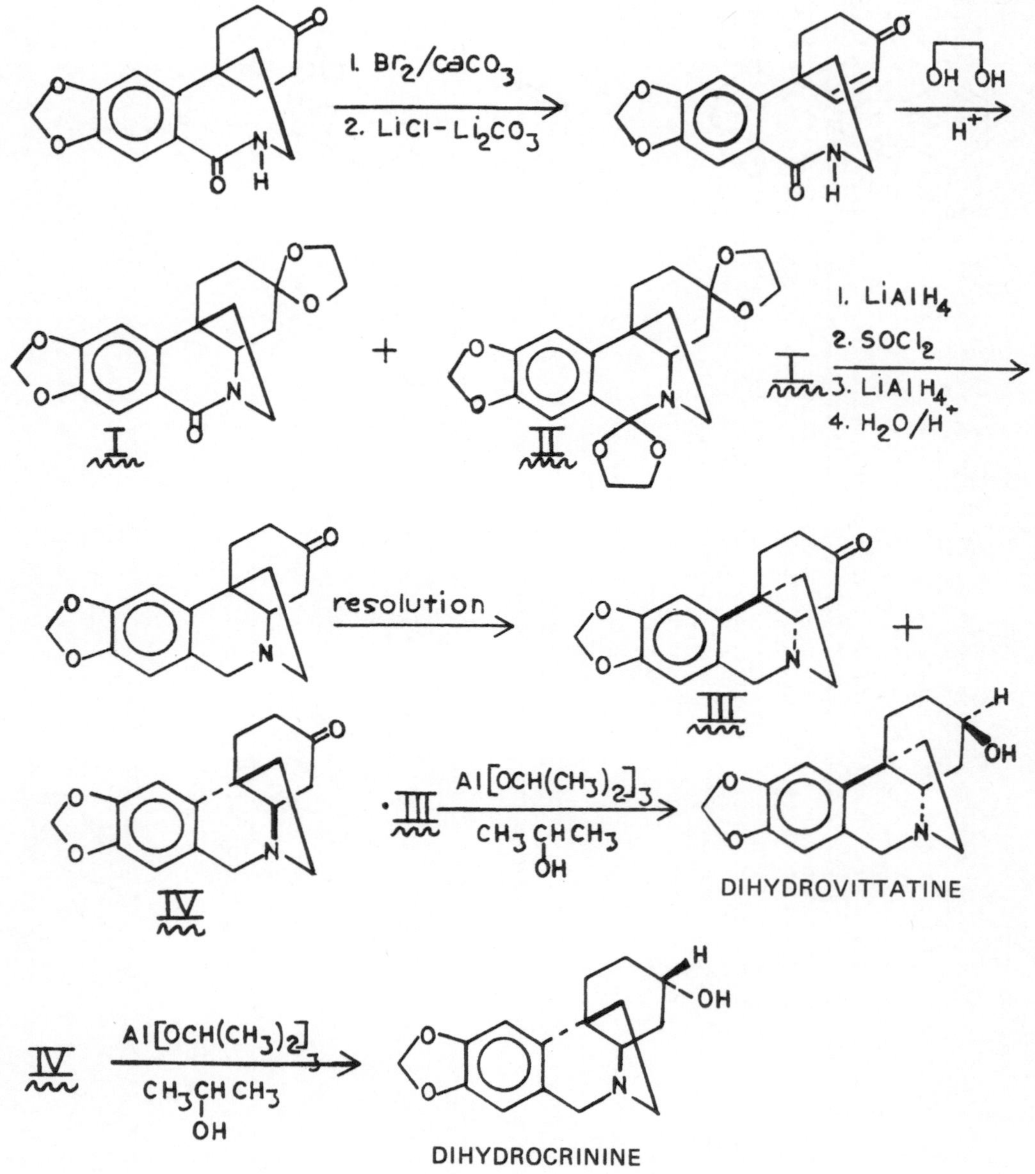

DIHYDROVITTATINE

DIHYDROCRININE

I-12   An intramolecular Michael reaction is employed to close the last ring of the crinine system. It is of interest to note the formation of an apparent lactam acetal (Structure II) under ketalization conditions. This anomaly vanishes when one notes that the usual amide resonance is eliminated when the B ring is formed. The bridgehead environment of the nitrogen, in the system so produced, mitigates against its assuming the sp^2 hybridization necessary for amide resonance. The ketonic rather than amidic nature of the carbonyl group in Structure I is also, no doubt, responsible for the complex series of steps ($LiAlH_4 - SOCl_2 - LiAlH_4$) necessary to reduce it to the corresponding methylene group.

I-13 DIHYDROGAMBIRTANNINE AND ASPIDOSPERMA-STRYCHNOS ALKALOID MODELS

E. Wenkert, K. G. Dave, C. T. Gnewuch, and P. W. Sprague; *J. Am. Chem. Soc.,* **90**, 5251 (1968).

DIHYDROGAMBIRTANNINE

ASPIDOSPERMA-STRYCHNOS MODEL

I-13   These syntheses are illustrative of the powerful new approach developed by the Wenkert school for constructing various indole alkaloids. The approach starts with N-alkylation of a nicotinic ester derivative with a 3β-haloethylindole system. The pyridinium structure so produced is selectively reduced to the stable β-aminoacrylate.

The first synthesis shows the application of this method to an indole containing a free α position. This allows for cyclization to a tetrahydrocarbazole system. The second case involves a system where the α position is blocked, thereby leading to cyclization at the β position (cf. I-24). After tautomerization to the exocyclic isomer (again a β-amino-acrylate), a tetracyclic model system of relevance to strychnos and aspidosperma alkaloids is isolated. Applications of this approach to nonindolic alkaloids are found in I-26.

For an excellent review of this approach see E. Wenkert, *Accts. Chem. Res.*, 1, 78 (1968).

I-14 15,16-DIMETHOXYERYTHRINAN-3-ONE

R. V. Stevens and M. P. Wentland; *Chem. Communs.*, 1104 (1968).

15,16-DIMETHOXYERYTHRINAN-3-ONE

I-14   The interesting point in this synthesis is a methyl vinyl ketone annelation reaction on the Δ^2-pyrroline system. It will be noted that a very similar reaction is employed in three syntheses of mesembrine (cf. I-30, I-31, and I-32). In all cases the azahydrindanone systems produced are *cis* fused. This could well reflect thermodynamic control over the last (Mannich) step.

I-15 SYNTHETIC APPROACHES TO DITERPENE ALKALOIDS

P. Grafen, H. J. Kabbe, O. Roos, G. D. Diana, T. Li, and R. B. Turner; *J. Am. Chem. Soc.,* **90**, 6131 (1968).

(B)

OMe — RCO_3H — OMe — ^-B — HO OMe

1. H_2/Pd
2. W K
3. CrO_3

H OMe — 1. Ac_2O 2. O_3 — OMe — BuONO $K^+ \ ^-O$+ —

HO N OMe — 1. TsCl 2. ^-OH — CN Me HO_2C H — $LiAlH_4$ —

H N O OMe — $LiAlH_4$ — H N OMe

I-15   These two approaches point the way to forging a -CH_2-N(Me)-CH_2- bridge between positions 4 and 10 (steroid numbering) required for the synthesis of diterpene alkaloids.

The first method starts with a rather ingenious preparation of a cyano diester by using 1,3-dibromopropane in a two-stage alkylation procedure (cf. I-30 and I-31). Catalytic reduction of the double bond and of the nitrile results in formation of a bridging lactam function. This synthesis avoids the serious issue of establishing the correct stereochemistry at C-4 by placing two carbomethoxyl groups at this position.

Synthesis I-15$_b$ comes to grips with this problem in a most clever fashion. Ring formation (via internal alkylation of an enone by an epoxide) and opening [second order Beckmann (cf. I-9)] ensures the required *cis* relationship between the carbonyl at C-4 and the cyano group at C-10. This allows for reductive lactam formation.

I-16 DITERPENE ALKALOID INTERMEDIATES

T. Matsumoto, M. Yanagiya, E. Kawakami, T. Okuno, M. Kakizawa, S. Yasuda, Y. Gama, J. Omi, and M. Matsunaga; *Tetrahedron Letters*, 1127 (1968).

$$ \xrightarrow[\text{2.K } \bar{O}+]{\text{1. NaIO}_4/\text{OsO}_4} $$

I

II

1. acetylation
2. SH SH/BF$_3$
3. Ni
4. LiAlH$_4$
5. CrO$_3$

$$ \xrightarrow[\text{K } \bar{O}+]{\text{I. (CH}_3)_2\text{CHCH}_2\text{CH}_2\text{ONO}} $$

2. SOCl$_2$

1. B$_2$H$_6$
2. LiAlH$_4$
3. acetylation

I-16 The strategy of ring forming-ring opening to establish the relative stereochemistry between two centers is quite similar to that employed in synthesis I-15$_b$. The ring is established through an aldol closure. It is interesting to note its occurrence even when β elimination of water is precluded (cf. I-27).

I-17   DITERPENE ALKALOIDS

A Stereospecific Synthesis of Pentacyclic Intermediates
with a Bridge in Ring B

K. Wiesner, A. Philipp, and P. Ho; *Tetrahedron Letters*, 1209 (1968).

I-17   The reader is urged to study the first paper in this series [see K. Wiesner and A. Philipp, *Tetrahedron Letters*, 1467 (1966)] which sets forth the preparation of the starting material and the general synthetic plan. A particularly noteworthy step is the rearrangement of the N-sulfonylaziridine to produce a sulfonamide in the appropriate *syn* relationship to the future A ring. The latter is closed through an intramolecular aldol condensation of a diketone. Conjugate addition of hydrogen cyanide affords a nitrile which leads to a lactam.

Another elegant feature of the synthesis is the lithium-ammonia deblocking of three (O-benzyl, O-acetyl, and N-sulfonyl) protecting groups in one synthetic operation.

I-18 EPIIBOGAMINE AND IBOGAMINE

W. Nagata, S. Hirai, T. Okumura, and K. Kawata; J. Am. Chem. Soc., **90**, 1650 (1968).

→ Methoxy lactam

1. LiAlH₄
2. alumina

$$\xrightarrow{}$$

EPIIBOGAMINE R_1 = H; R_2 = Et
IBOGAMINE R_1 = Et; R_2 = H

I-18   Epimer I_A is formed through a Claisen rearrangement on a precursor of known stereochemistry [cf. A.W. Burgstahler and I.C. Nandin, *J. Am. Chem. Soc.*, 83, 198 (1961)]. Epimer I_B apparently results from kinetically controlled (equatorial) α-carbon protonation of the hydrazone anion (cf. IIB-7).

Oxidative closure of an aziridine ring [cf. W. Nagata, S. Hirai, K. Kawata, and T. Aoki, *J. Am. Chem. Soc.*, 89, 5045 (1967)] sets the stage for acylative cleavage (cf. I-6 and I-10). The concluding steps are similar to those previously employed [cf. G. Büchi, D. L. Coffen, K. Kocsis, P. E. Sonnet, and F. E. Ziegler, *J. Am. Chem. Soc.*, 88, 3099 (1966)], with the striking exception that rearrangement of the isoquinuclidine appears to have been avoided (cf. I-45).

I-19 EPIIBOGAMINE

Y. Ban, T. Wakamatsu, Y. Fujimoto, and T. Oishi; *Tetrahedron Letters*, 3383 (1968).

EPIIBOGAMINE

I-19   This route, which relies on a selective Diels-Alder reaction of cross-conjugated isomer **A** relative to the extended conjugated isomer **B** is very similar in design to that of Büchi (see reference in I-18). The difficulties in the later stages of the Büchi synthesis are avoided through recource to a Fischer indole synthesis to forge the indole system already properly fused to the isoquinuclidine.

I-20 SYNTHESIS OF RING SYSTEMS RELATED TO HASUBANAN ALKALOIDS

M. Tomita, M. Kitano, and T. Ibuka; *Tetrahedron Letters*, 3391 (1968).

I-20   This is an interesting route to propellanes [cf. D. Ginsburg, *Accts. Chem. Res.*, 2, 121 (1969)] which could have general applicability. In the particular case, the two alkylations are directed to the same carbon due to benzylic stabilization of the corresponding enolate.

I-21 HERNADINE

K. Soh and F. Lahey; *Tetrahedron Letters*, 19 (1968).

HERNADINE

I-21 The course of the copper-induced Pschorr reaction is in contrast with the result of the thermal reaction encountered in Synthesis I-3. Possibly ring size is important in determining the destiny of the closure [see the synthesis of amurine, T. Kametani, K. Fukumoto, and T. Sugahara, *Tetrahedron Letters*, 5459 (1968)].

I-22 HOMOERYTHRINADIENONE

T. Kametani and K. Fukumoto; *J. Chem. Soc.*, (*C*) 2156 (1968).

HOMOERYTHRINADIENONE

I-22   This synthesis involves, as its key step, a Barton-type phenolic coupling reaction. It will be noted that the last step involves a twofold oxidation. The sequence of the two-ring forming steps is unclear. Two possibilities have been presented by Scott and co-workers [see J. E. Gervay, F. McCapra, T. Money, G. M. Sharma, and A. I. Scott, *Chem. Communs.*, 142 (1966)], who synthesized erythrinadienone by a similar strategy.

I-23 HOMOPROAPORPHINE-TYPE COMPOUNDS

T. Kametani, H. Yagi, F. Satoh, and K. Fukumoto; *J. Chem. Soc.*, (*C*) 271 (1968).

I-23   Again a Barton-type closure is the critical step. It is interesting to note that an oxide bridge does not close by Michael addition to the spirodienone [cf. D. H. R. Barton, *Hugo Miller Lecture, Proc. Chem. Soc.*, 293 (1963)] .

I-24 β,β-DISUBSTITUTED INDOLINE DERIVATIVES

T. Oishi, M. Nagai, and Y. Ban; *Tetrahedron Letters,* 491 (1968).

Intermediate for synthesis
of strychnos compounds

I-24   The use of an indoxyl system to direct β closure is one of the key features of this synthesis. Another is the use of an imino ether (cf. I-52) to labilize an amide toward intramolecular attack by a proximate enolate. The authors point out that this particular imino ether fails to condense except by an intramolecular route (cf. I-52).

The ketodioxolane of formyl acetone is an interesting reagent which has not hitherto been used.

I-25   17-ACETYL-5α-ETIOJERVA-12,14,16-TRIEN-3β-OL

W. S. Johnson, N. Cohen, E. R. Habicht, D.P.G. Hamon, G. P. Rizzi, and D. J. Faulkner; *Tetrahedron Letters*, 2829 (1968).

17-ACETYL-5α-ETIOJERVA-12,14,16-TRIEN-3β-OL

I-25   Johnson exploits the availability of the hydrochrysene starting material which served him so well in steroid total synthesis. Contraction of the C ring is achieved through oxidation of a double bond which is introduced through acetoxylation-dehydro-acetoxylation. The workers note that the relative thermodynamic stability of the B-C junction is a function of the A-B stereochemistry.

The title compound was previously [W. S. Johnson, J. M. Cox, D. W. Graham, and H. W. Whitlock, *J. Am. Chem. Soc.*, 89, 4524 (1967)] synthesized from Hagemann's ester and converted to veratramine [W. S. Johnson, H. A. P. deJongh, C. E. Coverdale, J. W. Scott, and U. Burckhardt, *ibid.*, 89, 4523 (1967)] and jervine [T. Masamune, M. Takasugi, A. Murai, and K. Kobayashi, *ibid.*, 89, 4521 (1967)].

I-26 LUPININE AND EPILUPININE AND THE HYDROLULOLIDINE AND HYDROJULOLIDINE RING SYSTEMS

E. Wenkert, K. G. Dave, and R. V. Stevens; *J. Am. Chem. Soc.,* **90,** 6177 (1968).

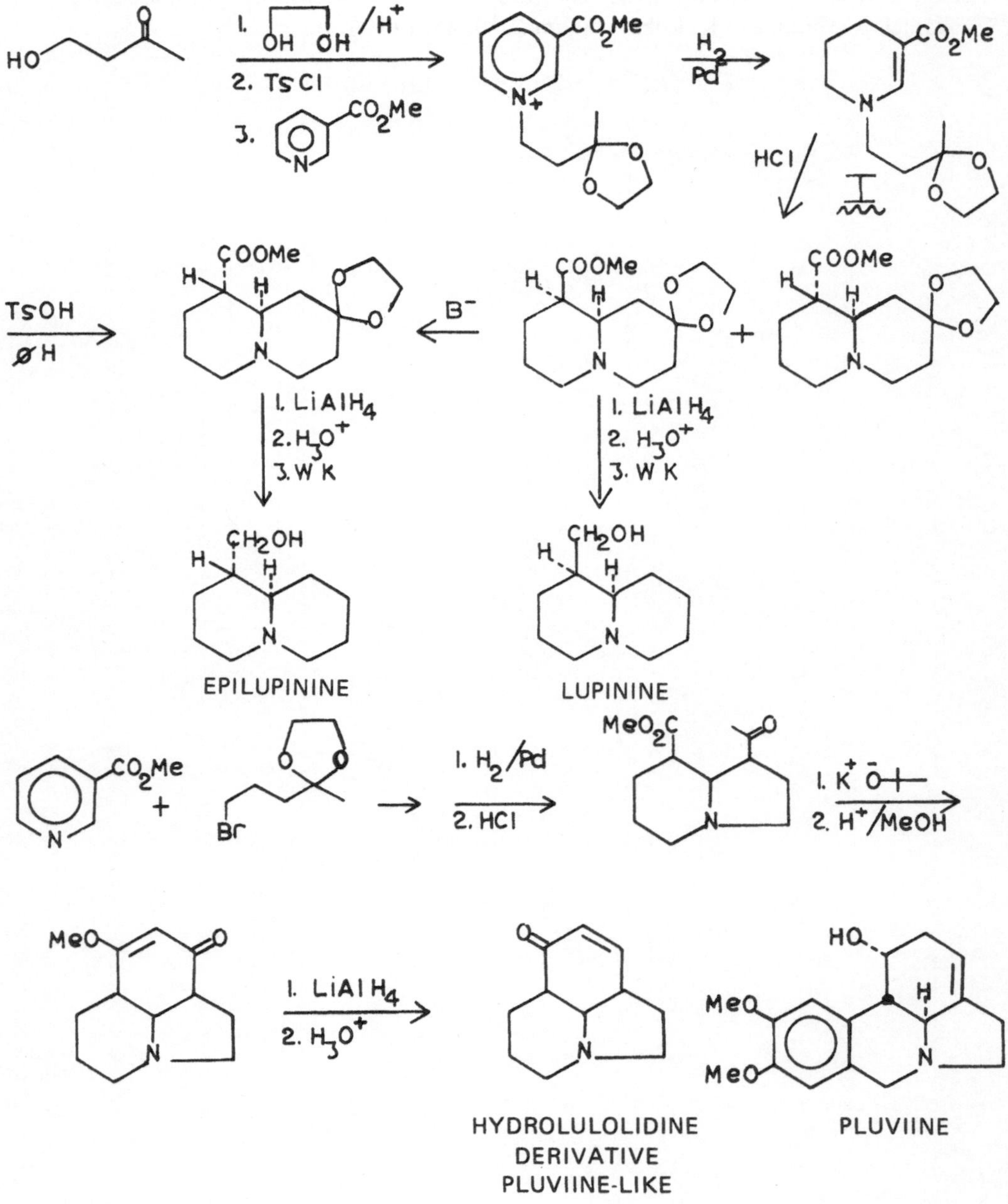

EPILUPININE

LUPININE

HYDROLULOLIDINE
DERIVATIVE
PLUVIINE-LIKE

PLUVIINE

PERHYDROJULOLIDONE   PERHYDROJULOLIDINE

I-26   In this paper we encounter three elegant applications of the Wenkert tetrahydropyridine approach to alkaloid synthesis (cf. I-13). Note should be taken of the use of a homologous series of terminally electrophilic ketals (cf. I-27 and IIIA-11).

Another interesting feature brought out in the three syntheses is the vulnerability of a β-aminoacrylate (a vinylogous amide) to intramolecular Michael addition. This type of reaction is also used in the synthesis of 12-epilycopodine which follows.

I-27 12-EPILYCOPODINE

K. Wiesner, V. Musil, and K. J. Wiesner; *Tetrahedron Letters*, 5643 (1968).

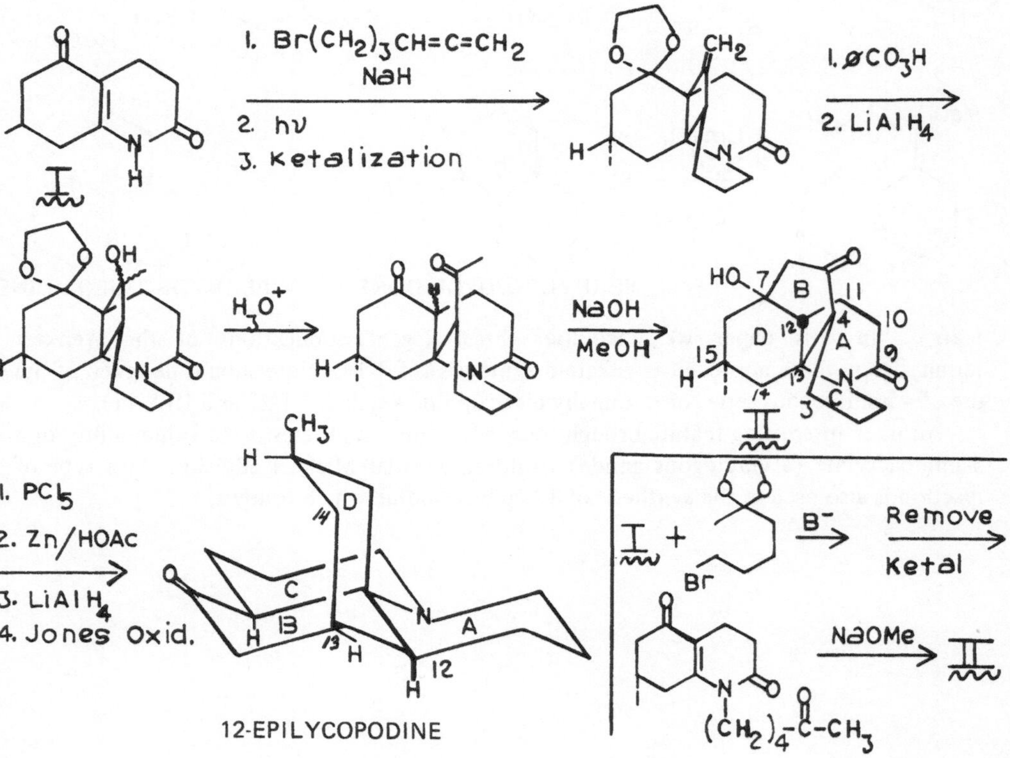

1. $Br(CH_2)_3CH=C=CH_2$
   NaH
2. $h\nu$
3. ketalization

1. ϕCO_3H
2. $LiAlH_4$

H_3O^+

NaOH
MeOH

1. PCl_5
2. $Zn/HOAc$
3. $LiAlH_4$
4. Jones Oxid.

12-EPILYCOPODINE

B^- Remove
        Ketal

$(CH_2)_4$—$\overset{O}{\overset{||}{C}}$—$CH_3$

NaOMe

I-28 LYCOPODINE

G. Stork, R. A. Kretchmer, and R. H. Schlessinger; *J. Am. Chem. Soc.,* **90**, 1647 (1968).

LYCOPODINE

I-29 LYCOPODINE

W. A. Ayer, W. R. Bowman, T. C. Joseph, and P. Smith; *J. Am. Chem. Soc.,* **90**, 1648 (1968).

Tetrahedron Letters, 2021 (1966).

LYCOPODINE

I-27, 28, 29   The numbering system and stereochemistry of lycopodine are shown below.

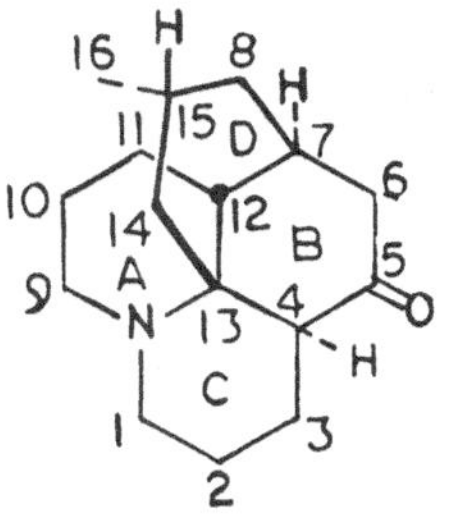

The stereochemical requirements which must be met in a synthesis of lycopodine are (i) a *trans* relationship between C-14 and the hydrogen at C-4, (ii) a *cis* relationship between the methyl group at C-15 and the hydrogen at C-7, and (iii) a *cis* relationship between C-14 and the hydrogen at C-12 (i.e., AB *cis* fused; AD *trans* fused). Of these, the first issue is easily resolved since the C-5 ketone allows for equilibration of the hydrogen at C-4 so as to afford the B-C *trans* fused system.

With these considerations in mind we turn first to synthesis I-27. Relationship (ii) is achieved in the first variant through a stereospecific photochemically induced cyclo-addition of an allene with the olefinic bond of an enamide. The bond between C-13 and C-14 is formed *trans* to the methyl. Since this bond must *per force* be *trans* (with respect to the future D ring) to the hydrogen at C-7, a *cis* relationship between the latter and the methyl is assured. The alternate synthesis establishes the C-13—C-14 bond *trans* to the methyl by an axial (cf. I-28) intramolecular Michael addition to the double bond of the vinylogous amide (cf. I-26). In both variants, the hydrogen at C-12 is introduced, in an apparently equilibrating step, *cis* to the C-4—C-13 bond and *trans* to the C-13—C-14 bond. This leads to a *trans* A-B, *cis* A-D mode of fusion which is the reverse of lycopodine.

Synthesis I-28 is the only one which is conducted with total control over the three stereochemical issues. Relationship (ii) is achieved through kinetically controlled (axial) conjugate addition of methyl Grignard to the enone. This leads to a *cis* relationship between the future C-7 hydrogen and methyl at C-15, which is maintained throughout the course of the synthesis. The *cis* A-B fusion, which satisfies requirement (iii), is established through a kinetically controlled *trans* addition of a proton and a *p*-anisyl group to the double bone of the enamide.

The use of trichloroethoxycarbonyl blocking group to protect an amine during the degredation of the anisole group should be noted.

A recent publication [see G. Stork, *Pure Appl. Chem.*, 17, 383 (1968)] sets forth an alternate procedure which bypasses the weakest point of this synthesis, the non-specificity in the enamine formation.

Synthesis I-29 is not stereospecific with respect to requirement (ii). Separation of epimers about the future C-15 is thus necessitated at an early stage. Requirement (iii) is assured through convex addition of the substituted Grignard to the starting immonium salt.

It is interesting to note that the oxidation of the amine to the lactam involves intermediate immonium unsaturation which is Δ^1 with respect to the *cis* fused A-B rings rather than the *trans* fused B-C system. Similarly, the observed intramolecular alkylation (cf. I-6) occurs via an enolate which is Δ^1 with respect to the *cis* fused A-B system and Δ^2 with respect to the *trans* fused (B-C) system rather than the reverse. These are in keeping with expected double bond stabilities in *cis-* and *trans*-decalins [cf. L. Velluz, J. Valls, and G. Nominé, *Angew. Chem. Int. Ed. Engl.*, 4, 181, (1965)].

I-30 MESEMBRINE

R. V. Stevens and M. P. Wentland; *J. Am. Chem. Soc.,* **90**, 5580 (1968).

MESEMBRINE

I-31 MESEMBRINE

S. L. Keeley and F. C. Tahk; *J. Am. Chem. Soc.*, **90**, 5584 (1968).

MESEMBRINE

I-32 MESEMBRINE

T. J. Curphey and H. L. Kim; *Tetrahedron Letters*, 1441 (1968).

MESEMBRINE

I-33 MESEMBRINE

T. Oh-ishi and H. Kugita; *Tetrahedron Letters*, 5445 (1968).

MESEMBRINE

I-34 MESEMBRINE

M. Shamma and H. R. Rodriguez; *Tetrahedron*, **24**, 6583 (1968).

I-30-34   These five syntheses of mesembrine illustrate several overlapping approaches. Syntheses I-30 and I-31 are obviously identical in design. They both use the cyclopropyl imine → Δ^2-pyrroline rearrangement (cf. I-4) and annelation of the latter with methyl vinyl ketone to produce a *cis* fused azahydrindanone (cf. I-14). Synthesis I-30 utilizes a preformed dilithio salt of a phenylacetonitrile.

Synthesis I-32 uses a different route to the Δ^2-pyrroline but culminates in the same annelation method. Synthesis I-33 utilizes a 1,1,1-trialkylated acetone in an interesting double cyclization to produce the appropriate azahydrindanone ring system. The *cis* and *trans* fused systems are eventually obtained.

Synthesis I-34 utilizes a 2-arylcyclohexanone to introduce an allyl group which is eventually modified so as to produce a β-aminoethyl grouping. The final ring closure may be viewed as a β elimination followed by a conjugate addition of nitrogen to the enone (cf. Synthesis I-12).

I-35 OCHOTENSINE AND OCHOTENSIMINE

H. Irie, T. Kishimoto, and S. Uyeo; *J. Chem. Soc.*, (*C*), 3051 (1968).

$\xrightarrow[\text{formalin}]{\text{HCl} \, \triangle}$

$\xrightarrow{}$

$\xrightarrow[\begin{array}{l}3.\ CH_2=O \diagup NaBH_4\end{array}]{\begin{array}{l}1.\ CH_2N_2\\2.\ \varnothing_3P=CH_2\end{array}}$

OCHOTENSIMINE

AISO

$\xrightarrow[\text{HCOOH}]{CH_2O}$

$\xrightarrow[2.\ ClCH_2OMe]{1.\ NaOH}$

$\xrightarrow[2.\ \text{Dil. HCl}]{1.\ \varnothing_3P=CH_2}$

OCHOTENSINE

I-36 OCHOTENSIMINE

S. McLean, M. S. Lin, and J. Whelan; *Tetrahedron Letters*, 2425 (1968).

OCHOTENSIMINE

I-35,36   The key step in these three closely related syntheses is the selective use of the β-carbonyl of a vicinal diketoindane system for Bischler-Napieralski interpolation into a dopamine-type residue (cf. I-7). The selectivity is presumably augmented by resonance between the α-ketone and the *para* oxygen function (see Synthesis IIA-3 for a similar argument).

I-37 PERLOLIDINE

J. C. Powers and I. Ponticello; *J. Am. Chem. Soc.*, **90**, 7102 (1968).

PERLOLIDINE

I-37   A delicate balance of acylating vs. nucleophilic capabilities of the malonic acid derivative governs the orientation of this pyridone synthesis [cf. C. R. Hauser and C. J. Eby, *J. Am. Chem. Soc.*, 79, 728 (1957)]. With an ester function, as in the case shown, the process appears to start with N-acylation and culminates in Knoevenagel closure.

I-38 PETALINE

G. Grethe, M. Uskoković, and A. Brossi; *J. Org. Chem.*, **33**, 2500 (1968).

OPT. ACTIVE
PETALINE $\longrightarrow$

PETALINE

I-38   The crux of this synthesis is the use of a benzyl ether blocking group to allow for (i) oxidative introduction of an immonium system via lead tetraacetate and (ii) Grignard addition. Having served its purpose, the benzyl group is removed by catalytic hydrogenolysis.

I-39 QUEBRACHAMINE SKELETON

F. E. Ziegler and P. A. Zoretic; *Tetrahedron Letters*, 2639 (1968).

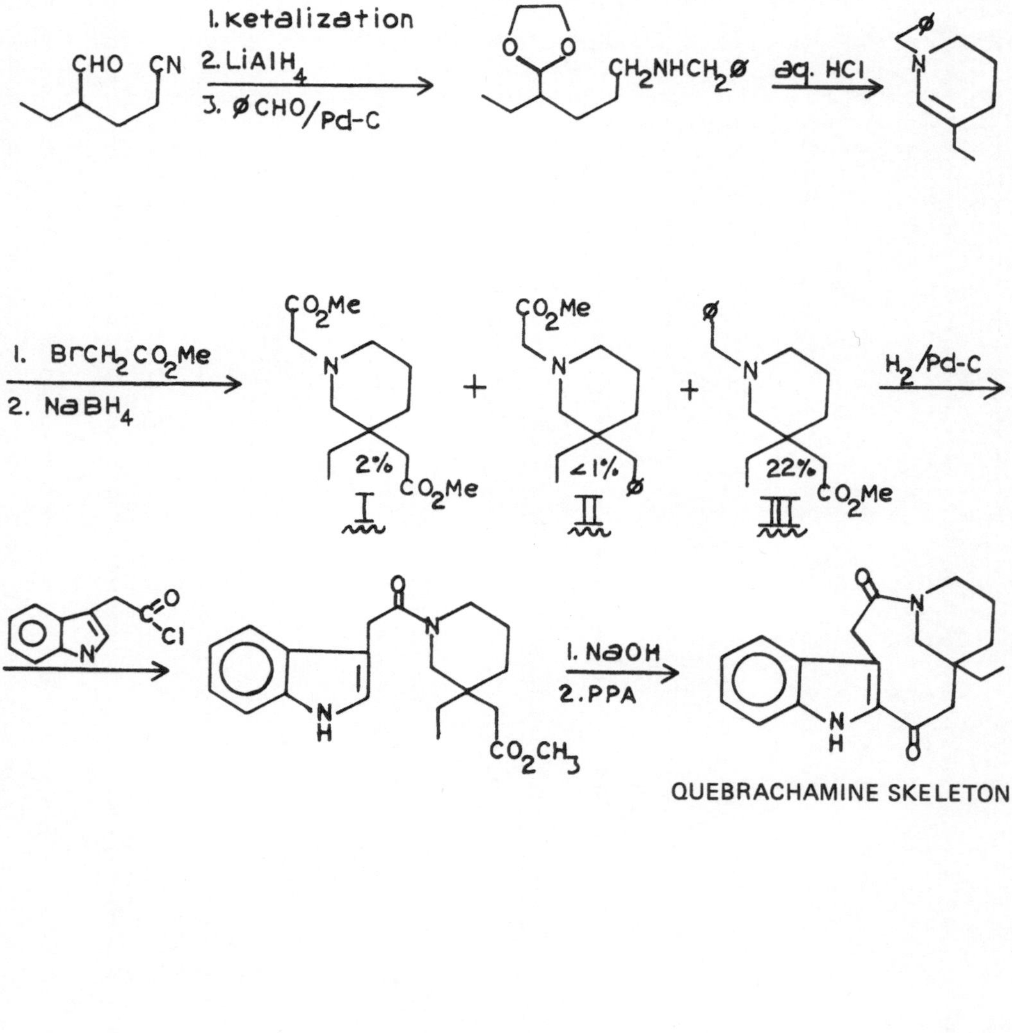
CHO
CN
1. Ketalization
2. LiAlH4
3. ØCHO/Pd-C
O
O
CH2NHCH2Ø
aq. HCl
Ø
N
1. BrCH2CO2Me
2. NaBH4
CO2Me
N
2%
CO2Me
I
+
CO2Me
N
<1%
Ø
II
+
Ø
N
22%
CO2Me
III
H2/Pd-C
O
Cl
N
N
H
CO2CH3
1. NaOH
2. PPA
O
N
N
H
O

QUEBRACHAMINE SKELETON

I-39 This elegant construction of the quebrachamine skeleton uses a preconstructed Δ^2-piperidine which is assembled through an intramolecular enamine formation. A required two-carbon connecting link is introduced by C-alkylation with ethyl bromoacetate. A β-indolylacetyl residue is introduced by acylation, and a nine-membered ring is closed through polyphosphoric acid cyclodehydration (cf. I-10 and I-50 for similar closures of six- and eight-membered rings).

I-40 SANGUINARINE

S. F. Dyke, B. J. Moon, and M. Sainsbury; *Tetrahedron Letters*, 3933 (1968).

SANGUINARINE

I-40 Reductive amination with aminoacetal sets the stage for an interesting condensation which may be formulated as shown:

I-41 SEMPERVIRINE HYDROBROMIDE

K. T. Potts and G. S. Mattingly; *J. Org. Chem.,* **33**, 3985 (1968).

SEMPERVIRINE
HYDROBROMIDE

I-41   This interesting synthesis of sempervirine is conceptually related to a previous synthesis [cf. R. B. Woodward and W. M. McLamore, *J. Am. Chem. Soc.*, 71, 379 (1949)]. In gross terms, sempervirine consists of harman + cyclohexanone + a one-carbon unit. In the Woodward synthesis, the one-carbon unit is attached to the cyclohexanone via formylation. In the Potts synthesis, a two-carbon unit is attached to harman by alkylation with ethyl bromoacetate. After condensation with cyclohexane-1,2-dione, the extraneous carboxyl is removed.

In principle, treatment of the carbethoxymethylated harman with base could give the ylid (proton loss from methylene) or the anhydro base (proton loss from methyl). It is interesting to note that under relatively mild conditions a product corresponding to condensation of the ylid with one carbonyl of the dione is isolated [cf. F. Kröhnke, *Angew. Chem. Int. Ed. Engl.*, 2, 225 (1963)].

I-42 SOLANIDINE

S. V. Kessar, R. K. Mahajan, S. S. Gandhi, and A. L. Rampal; *Tetrahedron Letters*, 1547 (1968).

I-42. Although the stereochemistry at C-25 is controlled through recourse to a resolved starting material (no racemization occurs at this center via enolization!) separation of isomers about C-20 and C-22 was required.

I-43 STEBISIMINE

T. Kametani, O. Kusama, and K. Fukumoto; *J. Chem. Soc.*, (*C*) 1798 (1968).

STEBISIMINE

I-43   The closure of a formal twenty six-membered *bis* lactam ring is the most noteworthy feature of this synthesis. The lack of orientational control over the mode of combination of the two unsymmetrical components (ca. equal amounts of **I** and **II** were isolated) is its major weakness. This is indeed a difficult problem in the synthesis of bisbenzylisoquinoline alkaloids of this type. For an alternate approach see Y. Inubashi and Y. Masaki, *Tetrahedron Letters*, 3399 (1968).

I-44 VEATCHINE

K. Wiesner, S. Uyeo, A. Philipp, and Z. Valenta; *Tetrahedron Letters*, 6279 (1968).

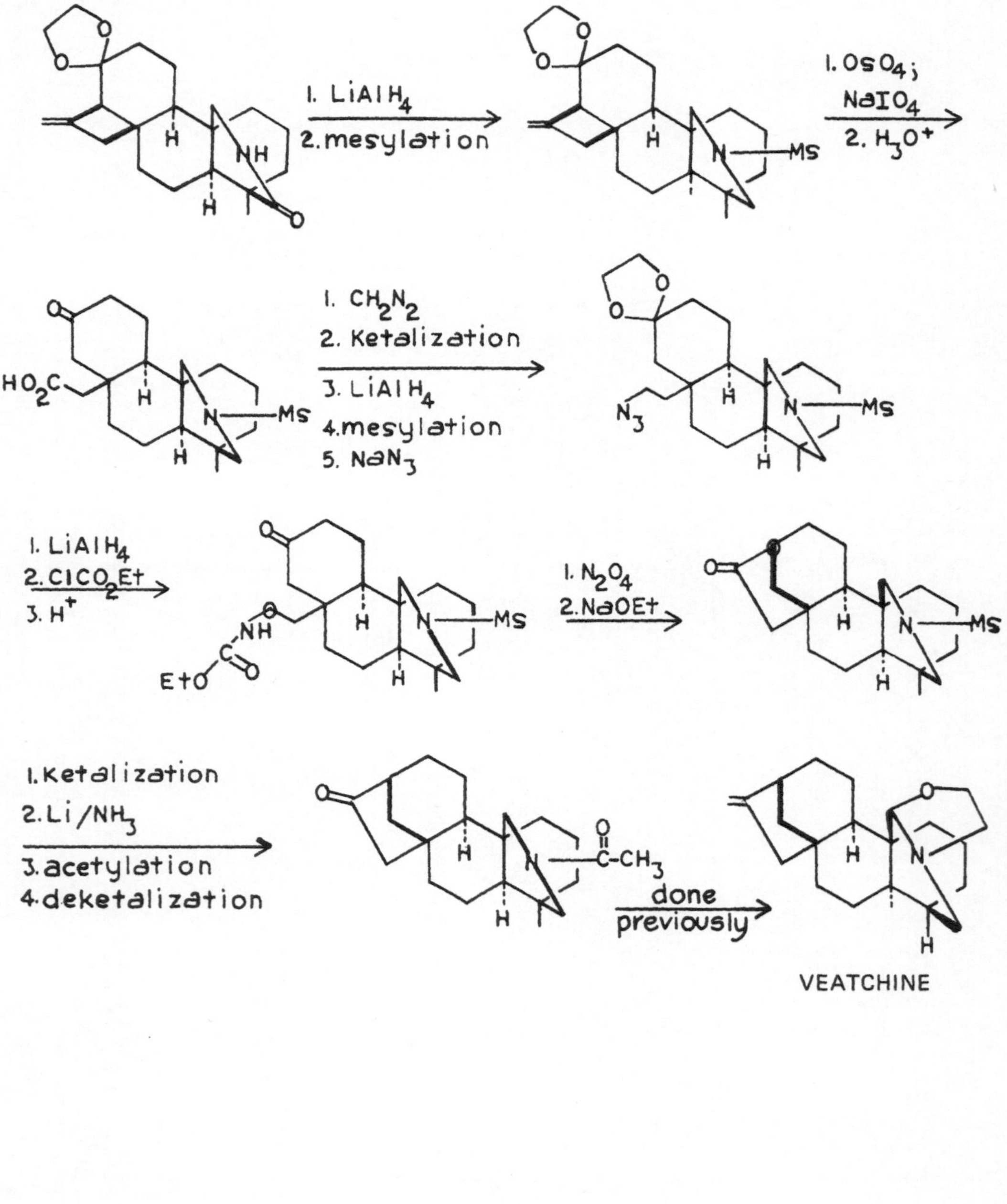

1. LiAlH$_4$
2. mesylation

1. OsO$_4$;
   NaIO$_4$
2. H$_3$O$^+$

1. CH$_2$N$_2$
2. Ketalization
3. LiAlH$_4$
4. mesylation
5. NaN$_3$

1. LiAlH$_4$
2. ClCO$_2$Et
3. H$^+$

1. N$_2$O$_4$
2. NaOEt

1. Ketalization
2. Li/NH$_3$
3. acetylation
4. deketalization

$\xrightarrow{\text{done previously}}$

VEATCHINE

I-44   The reader is urged to consult a previous paper [R. W. Guthrie, Z. Valenta, and K. Wiesner, *Tetrahedron Letters*, 4645 (1966)] for the synthesis of the starting material. An intermolecular photochemically induced allene-ethylene cycloaddition (cf. I-27) was used to close the four-membered ring.

A *retro* Dieckmann opening (cf. *retro* aldol opening in I-27) is used to generate a required ketoacid. The *crescendo* is reached when a diazoalkane, derived from the amine corresponding to the acid, effects an intramolecular ring expansion [cf. C. D. Gutsche and D. M. Bailey, *J. Org. Chem.*, 28, 607 (1963)] of the cyclohexanone giving a bicyclo[3,2,1] octane system of correct stereochemistry.

I-45 VELBANAMINE

G. Büchi, P. Kulsa, and R. L. Rosati, *J. Am. Chem. Soc.,* **90**, 2448 (1968).

VELBANAMINE

I-45   In this synthesis, Büchi and co-workers retrace the steps previously [G. Büchi, D. L. Coffen, K. Kocsis, P. E. Sonnet, and F. E. Ziegler, *J. Am. Chem. Soc.*, 88, 3099 (1966)] worked out in connection with the ibogamine problem (cf. I-18 and I-19). As in the Nagata work (cf. I-18), rearrangement is avoided during closure of the indole-isoquinuclidine bond. A *retro* aldol cleavage (cf. I-27) establishes the azabicyclododecane ring. Stereospecific addition of ethyl Grignard *cis* to the one-carbon bridge establishes the proper relative configuration at the one independent asymmetric center.

I-46 VERARINE

J. P. Kutney, J. Cable, W. A. F. Gladstone, H. W. Hanssen, E. J. Torupka, and W. D. C. Warnock; *J. Am. Chem. Soc.,* **90**, 5332 (1968).

VERARINE

I-46   The *anti* relationship between the hydrogens at C-8 and C-13 is established through thermodynamic control (via enolization) at the dienone stage. The *trans* B-C stereochemistry is established through catalytic reduction of the C-9—C-11 double bond. The latter is first introduced through an interesting cleavage of a β-acetoxyaldehyde (for a different way of removing the extraneous aldehyde see I-25).

I-47 MONOMERIC VINCA ALKALOIDS—VINCADINE, VINCAMINOREINE, VINCAMINORINE, VINCADIFFORMINE, AND MINOVINE

J. P. Kutney, K. K. Chan, A. Failli, J. M. Fromson, C. Gletsos, and V. R. Nelson; *J. Am. Chem. Soc.*, **90**, 3891 (1968).

I; R = H
II; R = CH_2-CH=CH_2
III; R = CH_2-CHO

$I + CH_2=CH CH_2 Br \xrightarrow{B^-} II \xrightarrow[NaIO_4]{OsO_4}$

$III \xrightarrow{} IV \xrightarrow{LiAlH_4}$

$V \xrightarrow[Pd/C]{H_2} VIA + VIB \xrightarrow{CH_3SO_2Cl}$

C_3-H and Et
are
cis in A
trans in B

IV; R = $CH_2\phi$, R_1 = O, R_2 = H, R_3 = H
V; R = $CH_2\phi$, R_1 = H_2, R_2 = H, R_3 = H
VI; R = R_1 = R_2 = R_3 = H

$VII \xrightarrow[NH_3]{Li}$

QUEBRACHAMINE: R= R_1=R_2=R_3= H
VIII; R = R_2 = R_3 = H, R_1 = CN
IX; R = R_1 = R_3 = H, R_2 = CN
X; R = R_2 = R_3 = H, R_1 = COOCH$_3$
XI; R = R_1 = R_3 = H, R_2 = COOCH$_3$
XII; R = CH_3, R_2 = H, R_1= COOCH$_3$ R_3 = H
XIII; R = CH_3, R_1 = H, R_2= COOCH$_3$,R_3 = H

QUEBRACHAMINE · VII $\xrightarrow{KCN}$ VIII +

$IX \xrightarrow[2. CH_2N_2]{1. OH^-} X$ +

VINCADINE

XI
EPIVINCADINE

· X $\xrightarrow{N-methylation}$

XII
VINCAMINOREINE

+

XIII
VINCAMINORINE

$X \xrightarrow[Pd/C]{O_2} XIV$
VINCADIFFORMINE

$XII \xrightarrow[Pd/C]{O_2} XV$
MINOVINE

XIV; R = H, R_3 = H
XV; R = CH_3, R_3 = H

I-47   The key steps in the synthesis of quebrachamine and related vinca alkaloids are formation of Structure **VII** via intramolecular quaterinization of the nitrogen and cleavage of the C-3—N_b bond (cf. acylative cleavage in I-6) either reductively or by cyanide displacement. This approach to the tetracyclic lactams is similar to that encountered in I-49. The transformation of **X** and **XII** to **XIV** and **XV**, respectively, is of a type encountered in the condyfoline-tubifoline paper (I-6). Tautomerization to the vinylogous urethan occurs in this case, and only one of two possible dehydro products is isolated (cf. I-6).

I-48 DIMERIC VINCA ALKALOIDS

J. P. Kutney, J. Beck, F. Bylsma, and W. J. Cretney; *J. Am. Chem. Soc.*, **90**, 4504 (1968).

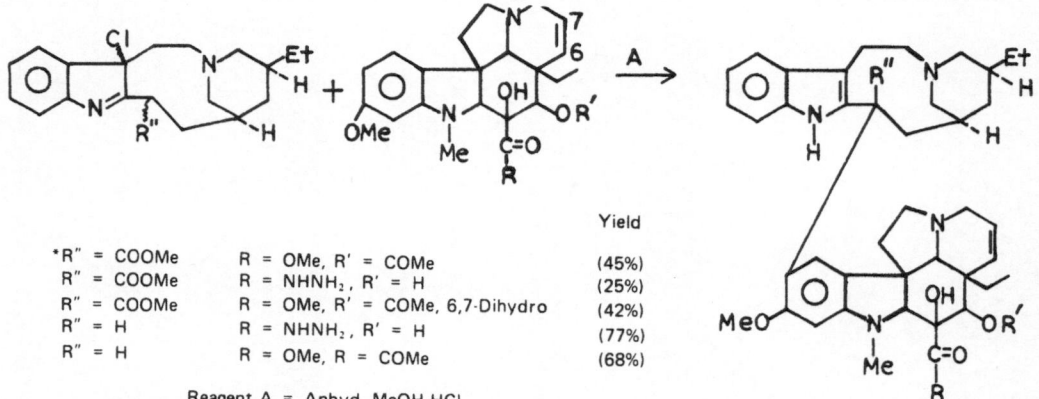

		Yield
*R″ = COOMe	R = OMe, R′ = COMe	(45%)
R″ = COOMe	R = NHNH₂, R′ = H	(25%)
R″ = COOMe	R = OMe, R′ = COMe, 6,7-Dihydro	(42%)
R″ = H	R = NHNH₂, R′ = H	(77%)
R″ = H	R = OMe, R = COMe	(68%)

Reagent A = Anhyd. MeOH-HCl

*Product is vindoline

I-48   The mechanism of this coupling reaction, which is of potential importance in the synthesis of antitumor dimeric vinca alkaloids, may be qualitatively formulated as follows:

⟶ Product

The tautomerization of the indolenine to the vinylogous urethan was encountered in I-13 and in the previous synthesis.

I-49 VINCAMINOL

J. Harley-Mason, L. Castedo, and T. J. Leeney; *Chem. Communs.*, 1186 (1968).

VINCAMINOL

I-49   The approach to the tetracyclic lactam is not unlike that encountered in I-47. As in that case, epimers form about C-3.

I-50 1-METHYL-16-DEMETHOXYCARBONYL-20-DESETHYLIDENE-VOBASINE

T. Shioiri and S. Yamada; *Tetrahedron*, **24**, 4159 (1968).

Title compound which
has vobasine skeleton

I-50   This synthesis is interesting in that the carbonyl carbon of tryptophan (methyl ester) is incorporated in a required piperidone ring. In its overall design, this synthesis is not unlike those encountered in I-10 and I-39.

I-51 WITHASOMNINE

A. Morimoto, K. Noda, T. Watanabe, and H. Takasugi; *Tetrahedron Letters*, 5707 (1968).

Method A

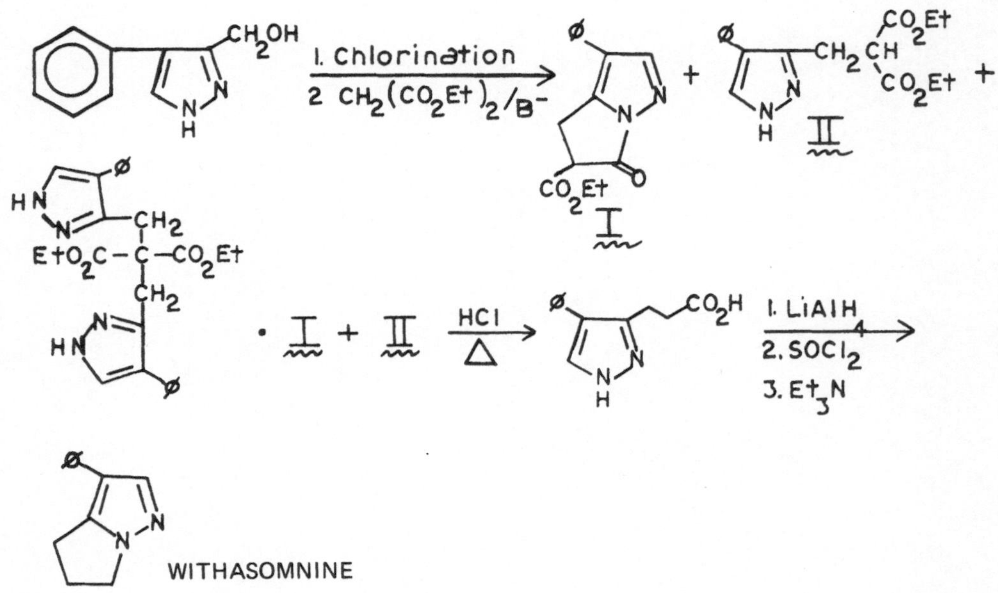

Method B

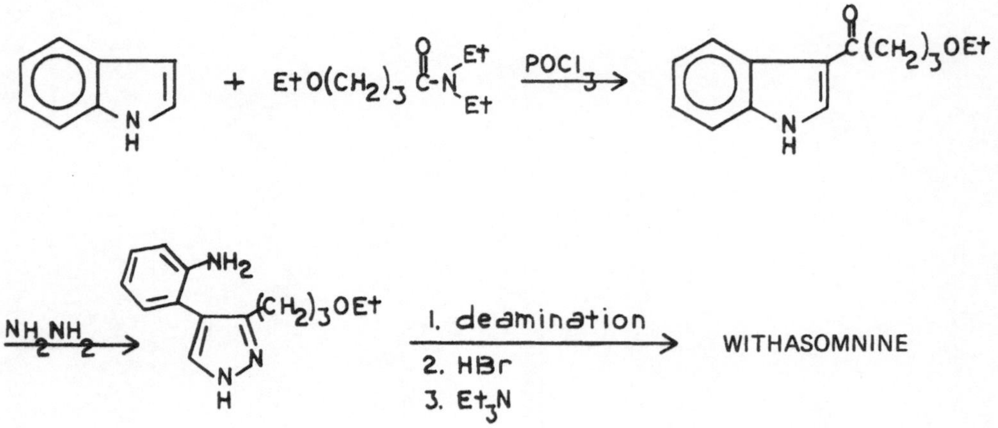

I-52 WITHASOMNINE

T. Onaka; *Tetrahedron Letters*, 5711 (1968).

WITHASOMNINE

I-51, 52   It is interesting to analyze the three different approaches to withasomnine. The first approach in I-51 uses a preformed phenyl hydroxymethylpyrazole (formed from 1-phenyl, 1-formyl acetone followed by oxidation). A malonic ester synthesis gives the required two-carbon extension which sets the stage for intramolecular N-alkylation (I-47).

The second approach in I-51 is fundamentally the same except that 3ω-ethoxy-butyryl indole serves as the progenitor of the pyrazole. The latter approach requires an awkward deamination of an extraneous amino group but has the advantage of establishing the three-carbon unit (of appropriate oxidation level) in a more satisfying fashion.

The approach in I-52 is the most elegant in that it is conceived along biogenetic lines and executed with beautiful simplicity. The use of an unhindered imino ether (cf. I-24) as an ornithine equivalent is a key feature of translating the conception to laboratory reality.

CHAPTER TWO

Terpenes and
Related Structures

IIA-1 ACTINIDIOLIDE, DIHYDROACTINIDIOLIDE, AND ACTINIDOL

S. Isoe, S. B. Hyeon, H. Ichikawa, S. Katsumura, and T. Sakan; *Tetrahedron Letters*, 5561 (1968).

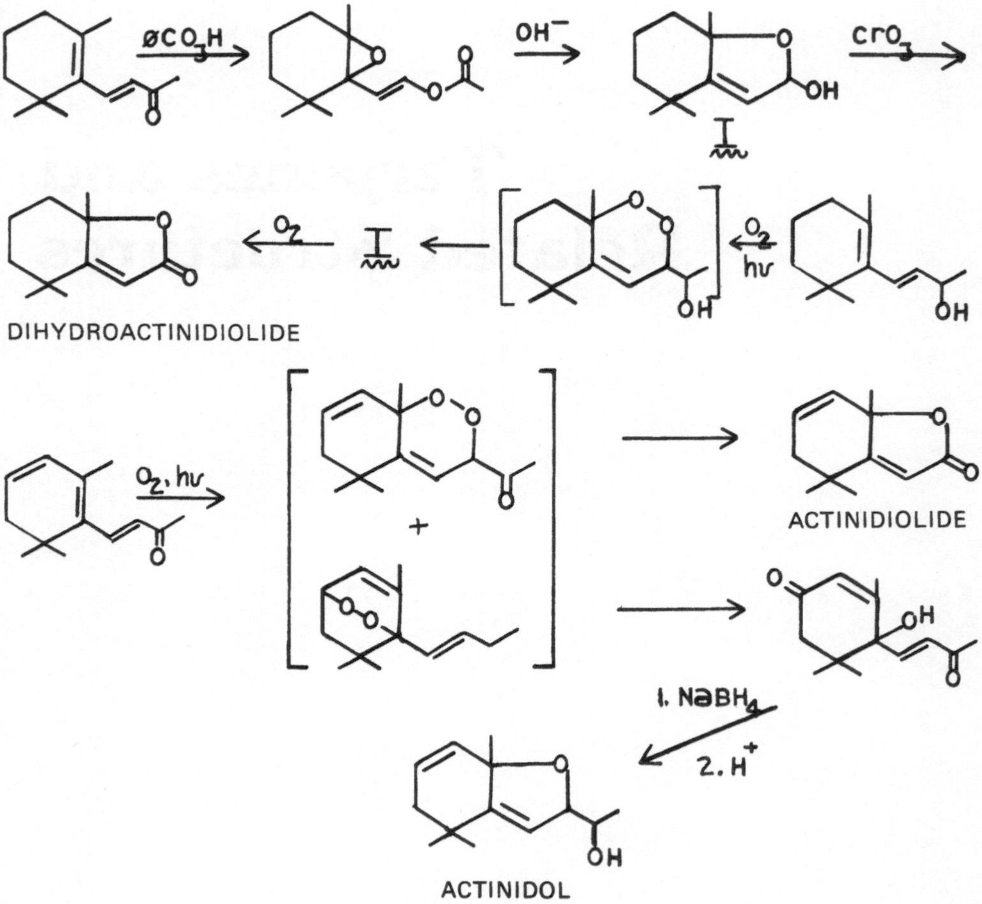

DIHYDROACTINIDIOLIDE

ACTINIDIOLIDE

ACTINIDOL

IIA-1   The key step in the synthesis of dihydroactinidiolide is the first one where
β-ionone reacts with 2 moles of perbenzoic acid to give an epoxyenolacetate.

The reader is urged to consult the original paper for some interesting mechanistic
proposals to account for the less obvious transformations of the cyclic peroxides.

IIA-2 NEZUKONE

A. J. Birch and R. Keeton; *J. Chem. Soc.*, (*C*) 109 (1968).

NEZUKONE

IIA-2   The Birch reduction-ring expansion sequence is a well-known route to a tropone ring such as embodied in nezukone. The dichlorocarbene addition gave 80% of monoadduct and 20% of a presumed diadduct.

IIA-3 ROSEFURAN AND DEHYDROELSHOLTZIONE

G. Büchi, E. Kovats, P. Enggist, and G. Uhde, *J. Org. Chem.*, 33, 1227 (1968).

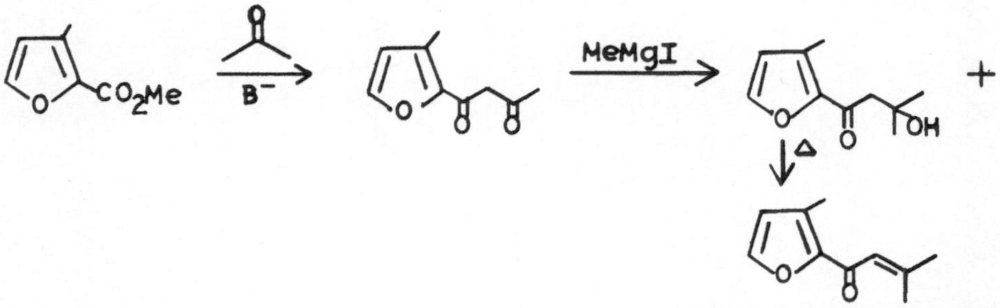

DEHYDROELSHOLTZIONE

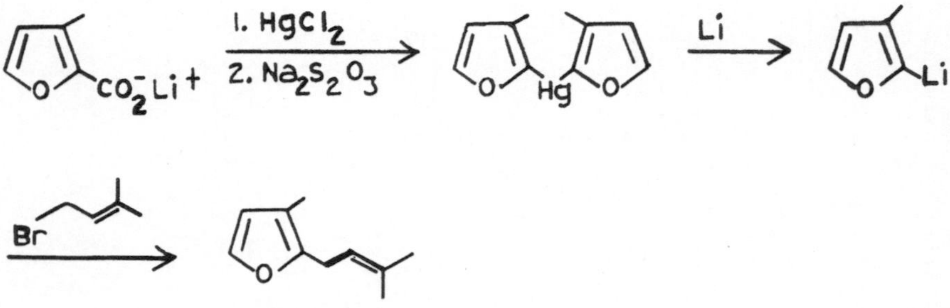

ROSEFURAN

IIA-3   The crucial step in the synthesis of dehydroelsholtzione is the addition of methyl Grignard to the unsymmetrical β-diketone. Two moles of the organometallic reagent are used (the first forms the magnesioenolate salt). It is interesting to note that the benzylic ketone is untouched (cf. I-35 and I-36) in the reaction. This may be the result of electronic deactivation of this carbonyl group by resonance with the furan oxygen, in addition to steric hindrance from the *ortho* methyl group.

The preparation of 2-lithio-3-methylfuran from 3-methyl-2-furoic acid via the mercury derivative should also be noted.

IIA-4 VERBENALOL

T. Sakan and K. Abe; *Tetrahedron Letters*, 2471 (1968).

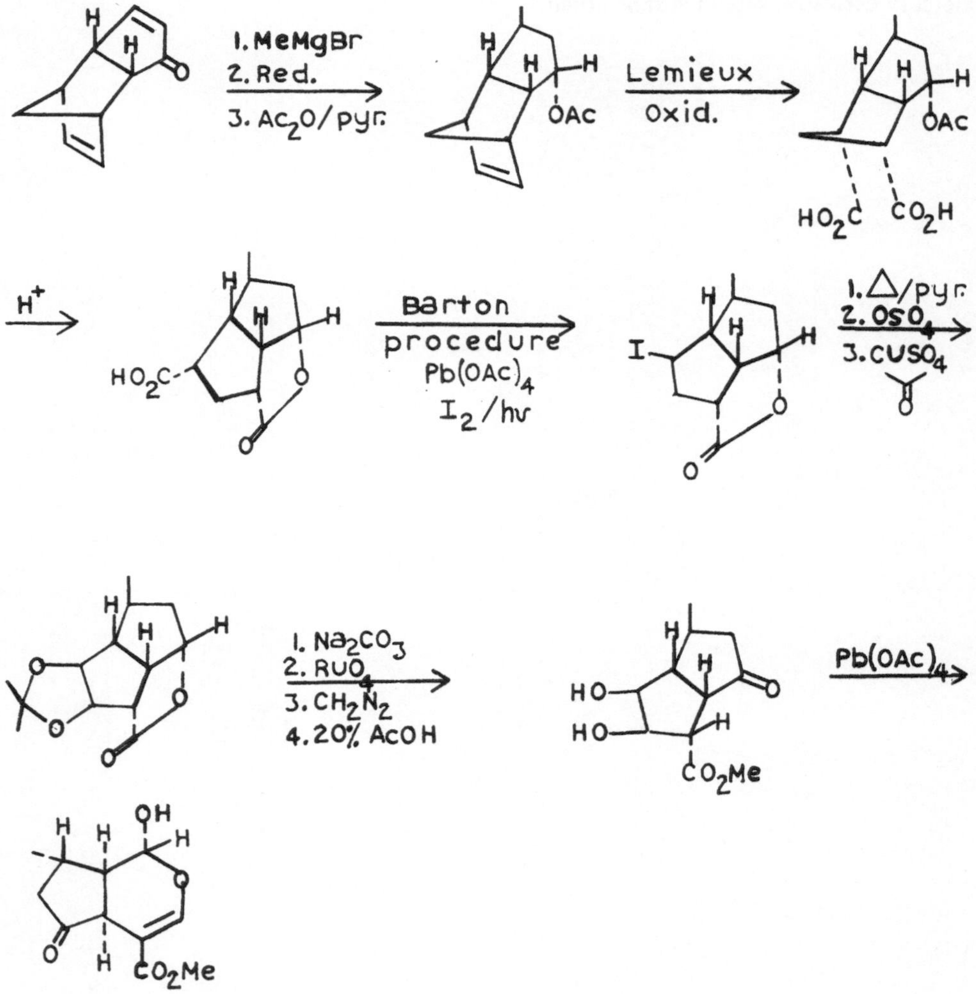
1. MeMgBr
2. Red.
3. Ac2O/pyr.
OAc
Lemieux
oxid.
OAc
HO2C
CO2H
H+
HO2C
Barton
procedure
Pb(OAc)4
I2/hν
I
1. △/pyr.
2. OsO4
3. CuSO4
1. Na2CO3
2. RuO4
3. CH2N2
4. 20% AcOH
HO
HO
CO2Me
Pb(OAc)4
OH
CO2Me

VERBENALOL

IIA-4   The *cis* ring junction in verbenalol is established from dicyclopentadienone. The *cis* relationship of the methyl and the junction hydrogens is arranged by a conjugate 1,4 addition of methyl Grignard from the less-hindered side of the molecule.

The penultimate dihydroxyketo ester is cleaved to a dialdehyde which suffers cyclization to the enol lactol. Thermodynamic control at the anomeric carbon gives the convex hydroxyl group.

IIA-5 SABINENE, SABINA KETONE, AND SABINENE HYDRATE

W. I. Fanta and W. F. Erman; *J. Org. Chem.*, **33**, 1656 (1968).

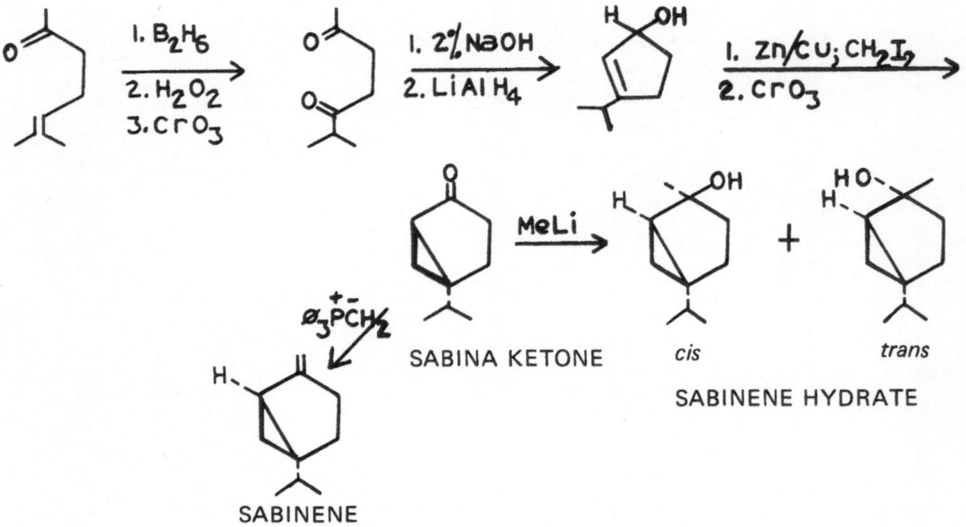

SABINA KETONE

cis   *trans*

SABINENE HYDRATE

SABINENE

IIA-5   The intramolecular aldol closure of 6-methyl-2,5-heptanedione occurs so as to result in dehydration to the α,β-unsaturated enone. This dehydration appears to govern the direction of the reaction.

It is found to be advantageous to conduct the Simmons-Smith reaction at the allylic alcohol stage. The cyclopropyl carbinol is then oxidized to sabina ketone.

IIB-1 ABSCISIC ACID

D. Roberts, R. Heckman, B. Hege, and S. Bellin; *J. Org. Chem.*, **33**, 3566 (1968).

ABSCISIC ACID

IIB-1   The key step in the synthesis of abscisic acid is *t*-butyl chromate oxidation of α-ionone to the hydroxydiketo derivative (23 to 27% yield). Selectivity in the Wittig reaction with respect to the two ketones should be noted.

IIB-2 JUVABIONE

B. Pawson, H. Cheung, S. Gurbaxani, and G. Saucy; *Chem. Communs.*, 1057 (1968).

JUVABIONE

IIB-2 The course of oxidation of the methylcyclohexene is apparently as follows:

Product

IIB-3 ANDROGRAPHOLIDE LACTONE

S. W. Pelletier, R. L. Chappell, and S. Prabhakar; *J. Am. Chem. Soc.*, **90**, 2889 (1968).

ANDROGRAPHOLIDE LACTONE

IIB-3   One of the interesting features of this synthesis is the use of the Nazarov reagent (methyl acrylylacetate) in an annelation reaction to give the Δ^4-3-keto-4-ester. This method has been popularized by Wenkert [see E. Wenkert, A. Afonso, J. B. Bredenberg, C. Kaneko, and A. Tahara, *J. Am. Chem. Soc.*, 86, 2038 (1964)].

Methylation of the enolate derived from the Δ^4-3-keto-4-carbomethoxy bicyclic system occurs in an equatorial sense (see IIC-5 and IIIC-1 for a similar result on similar substrates). The *trans* A-B stereochemistry is achieved by catalytic reduction of the 4,4-dialkylated-5,5-dehydro system [cf. G. Stork and J. W. Schulenberg, *J. Am. Chem. Soc.*, 84, 284 (1962)].

Elaboration of the butenolide follows the Sondheimer approach in the acetogenin series [see N. Danieli, Y. Mazur, and F. Sondheimer, *Tetrahedron*, 23, 715 (1967)]. The β configuration of the oxygen at C-2 arises from hydroxylation on the opposite side from the angular methyl.

IIB-4 AR-HIMACHALENE

R. C. Pandey and S. Dev; *Tetrahedron*, **24**, 3829 (1968).

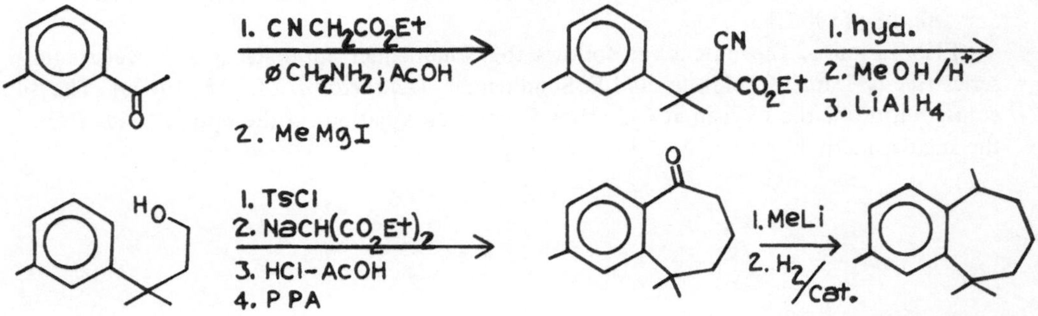

1. C N CH₂ CO₂Et

   $\overline{\emptyset\,CH_2NH_2\,;\,AcOH}$

2. Me Mg I

1. hyd.
2. MeOH/H⁺
3. LiAlH₄

1. TsCl
2. NaCH(CO₂Et)₂
3. HCl-AcOH
4. PPA

1. MeLi
2. H₂/cat.

AR-HIMACHALENE

IIB-4   Two-carbon chain extensions with malonate derivatives are conducted twice. The first extension employs a Knoevenagel reaction followed by conjugate Grignard addition to the activated benzylidene cyanoacetate double bond. It is interesting to note that copper catalysis was detrimental in this case, leading to reduction of the double bond. The addition went smoothly in the absence of copper catalysis.

IIB-5 BULNESOL

J. A. Marshall and J. J. Partridge; *J. Am. Chem. Soc.*, **90**, 1090 (1968).

70:30 ratio
of epimers.
α-CO₂Me more
stable.

BULNESOL

IIB-5   The key step in elaborating the hydrazulene skeleton of bulnesol is a planned rearrangement of a mesylate. The *trans* co-planar relationship of the departing mesylate with the migrating bond becomes apparent on examination of the structure below.

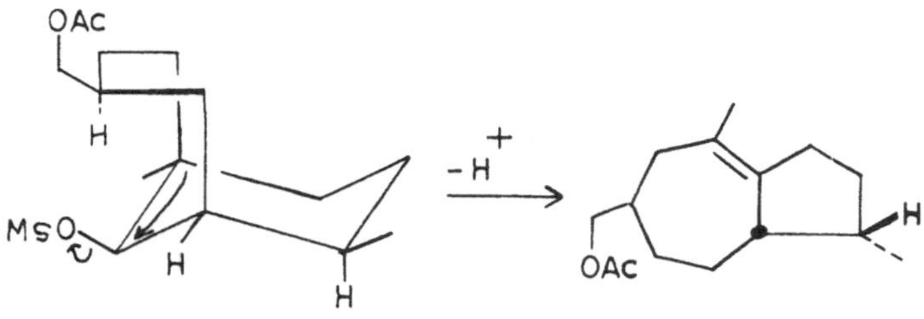

It will be noted that tertiary carbonium ion character is apparently influential in determining the course of the reaction.

The *trans* stereochemical relationship between the methyl group in the five-membered ring and the adjacent hydrogen was achieved by the entry of hydrogen on the least-hindered (*cis* to the 1-carbon bridge) side during catalytic reduction of the unsaturated keto ester.

The use of the novel *p*-chlorophenyl blocking group which is removed after (Birch) enol ether formation is another important feature of this remarkably beautiful synthesis.

IIB-6 CARABRONE

H. Minato and I. Horibe; *J. Chem. Soc.*, (*C*) 2131 (1968).

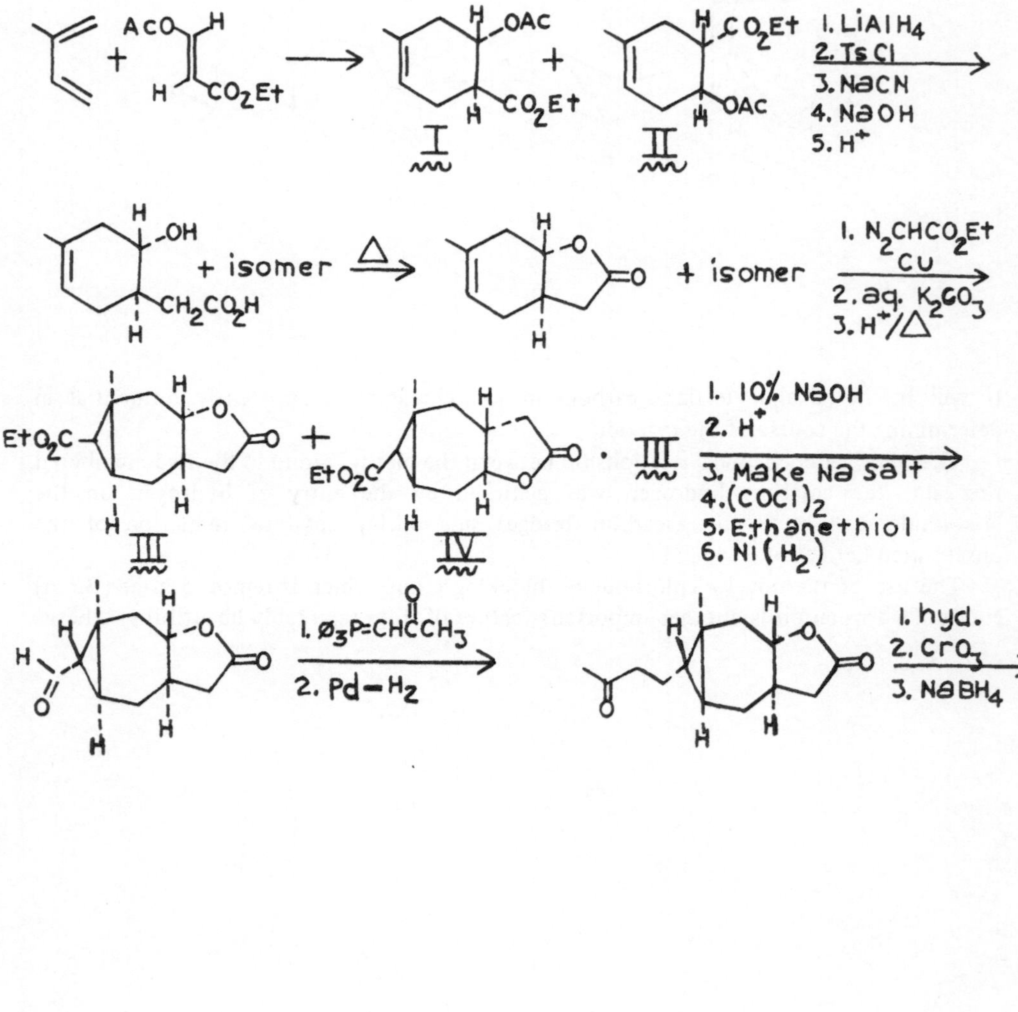

1. LiAlH$_4$
2. TsCl
3. NaCN
4. NaOH
5. H$^+$

1. N$_2$CHCO$_2$Et
   Cu
2. aq. K$_2$CO$_3$
3. H$^+$/$\triangle$

1. 10% NaOH
2. H$^+$
3. Make Na salt
4. (COCl)$_2$
5. Ethanethiol
6. Ni(H$_2$)

1. $\emptyset_3$P=CHCCH$_3$
2. Pd–H$_2$

1. hyd.
2. CrO$_3$
3. NaBH$_4$

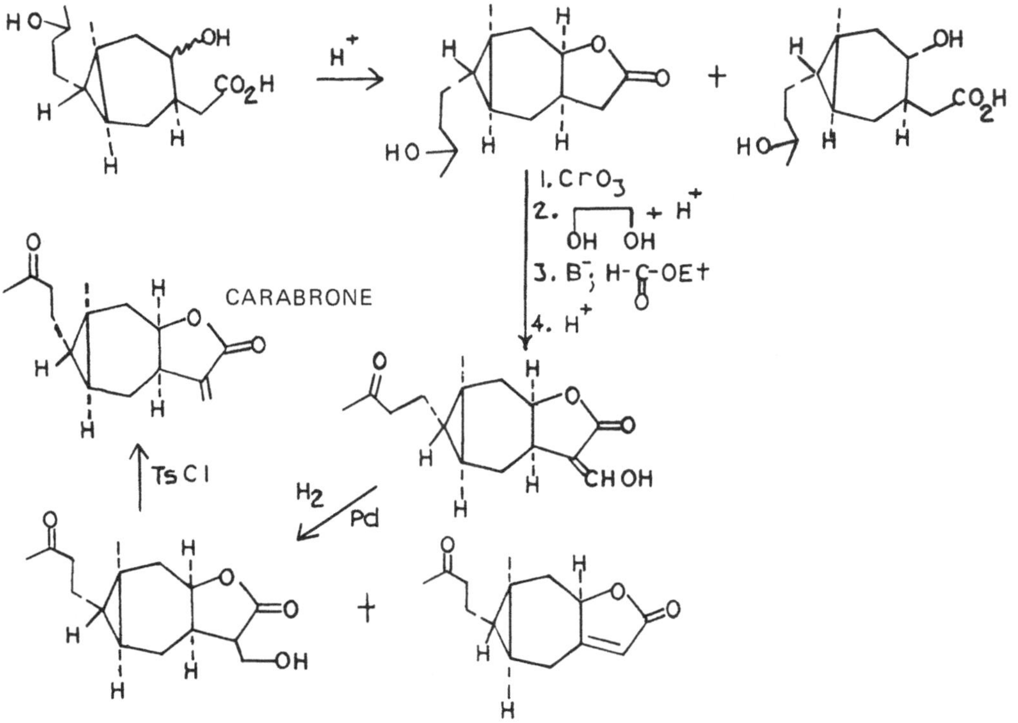

IIB-6   The orientational nonspecificity in the Diels-Alder reaction (3:1 in the desired sense) plagues this otherwise clever synthesis of carabrone. A *trans* lactone is established after suitable homologation of the adduct formed from β-acetoxyacrolein. This *trans* arrangement is quite important in directing the ethyldiazoacetate insertion reaction such that the cyclopropane ring forms *cis* to the methylene group adjacent to the lactonic carbonyl. Such a reaction on a *cis* fused system such as shown below would be expected to occur in a convex sense, producing the wrong stereochemistry required for carabrone.

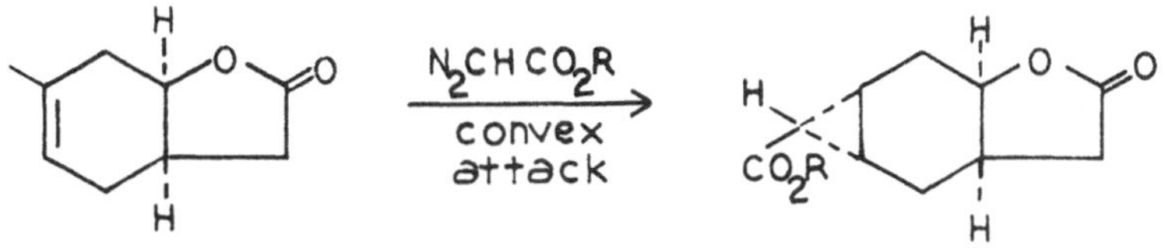

Having served its stereochemical purpose, the *trans* lactone ring is converted to the more stable *cis* system by exercizing thermodynamic control over the lactonization of the hydroxy acid intermediate.

The route to the α-methylene lactone had been previously developed by these workers [see H. Minato and I. Horibe, *J. Chem. Soc.*, (*C*) 1575 (1967)].

IIB-7 EREMOPHILENE AND EREMOLIGENOL

R. M. Coates and J. E. Shaw; *Tetrahedron Letters*, 5405 (1968).

EREMOLIGENOL   EREMOPHILENE

IIB-7   The preparation of the starting octalin is shown in the synthesis of aristolene (IIB-19). The crux of this synthesis is a twofold reduction of a β-acetoxyacrylate system with lithium in liquid ammonia. Kinetically controlled equatorial protonation (cf. I-18) gives rise to the axial carbomethoxy group necessary to give eremoligenol and then eremophilene.

IIB-8 EREMOPHIL-3,11-DIENE

E. Piers and R. J. Keziere; *Tetrahedron Letters*, 583 (1968).

EREMOPHIL-3,11-DIENE

IIB-8   The *cis* relationship between the isopropenyl group and the hydrogen at C-10 is controlled at the octalone stage through epimerization of the vinylogously activated bridgehead. This *cis* relationship allows for an equatorial disposition of the isopropenyl group. This should be compared with the case in IIB-15 where such an epimerization was not possible. This resulted in an axial isopropenyl group (*trans* to the junction methyl) in the major product.

The *cis* A-B fusion is achieved via lithium dimethyl copper addition to the enone.

A paper published subsequent to this report [R. E. Ireland and G. Pfister, *Tetrahedron Letters*, 2145 (1969)] suggests a simpler deoxygenation technique than was employed here. The Ireland approach also involves addition of lithium dimethyl cuprate to generate a *cis* fused lithium enolate. The latter is quenched by O-phosphorylation to give an enol phosphate which is subjected to reductive cleavage of the carbon-oxygen bond.

IIB-9 TETRAHYDROEREMOPHILONE

S. Murayama, D. Chan, and M. Brown; *Tetrahedron Letters*, 3715 (1968).

I A (R=CHO)   II A (R=CHO)

TETRAHYDROEREMOPHILONE

IIB-9   This synthesis is similar in design to Johnson's synthesis of fichtelite (IIC-4). The modified Hagemann's ester is prepared starting with methylacetoacetate and isobutyraldehyde. The desired I_A containing the axial isopropenyl group is the minor product ($I_A:II_A=2:3$) but is present in sufficient concentration to have suggested, with appropriate modification, an attractive approach to the elusive eremophilone. Unfortunately, dehydration of the α-ketol, necessary to introduce the enone system of eremophilone, did not occur. The difficulty in this reaction is presumably a reluctance of the ketol to adopt the conformation (in which the isopropyl group is axial) required to form the enone.

The photosensitized oxidation of the allylic methyl system was of a type encountered in IIB-2.

IIB-10 β-EUDESMOL

C. H. Heathcock and T. R. Kelly; *Tetrahedron*, **24**, 1801 (1968).

β-EUDESMOL

IIB-10   The key points in this synthesis are that the stereochemistry of the isopropyl carbinol and the bridgehead hydrogen are each controlled through precursors which allow for thermodynamic control.

IIB-11 DEHYDROFUROPELARGONES

G. Büchi and H. Wüest; *Tetrahedron*, **24**, 2049 (1968).

DEHYDROFUROPELARGONES

IIB-11   A Darzens glycidic ester condensation sets the stage for an interesting furan synthesis [cf. D. M. Burness, *J. Org. Chem.*, 21, 102 (1956)] involving an epoxyacetal. The mechanism may well be as follows:

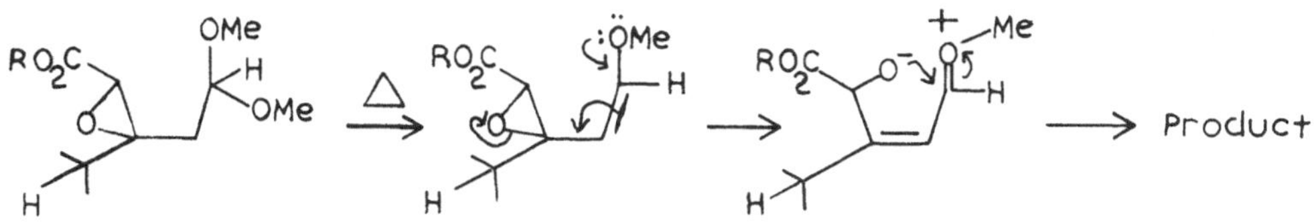

Again it will be noted that aldol closure on an unsymmetrical diketone occurs in the manner which allows for *retro* Michael elimination of water (cf. IIA-5).

IIB-12 ISOIRESIN, DIHYDROIRESIN, AND ISODIHYDROIRESIN

S. W. Pelletier and S. Prabhakar; *J. Am. Chem. Soc.*, **90**, 5318 (1968).

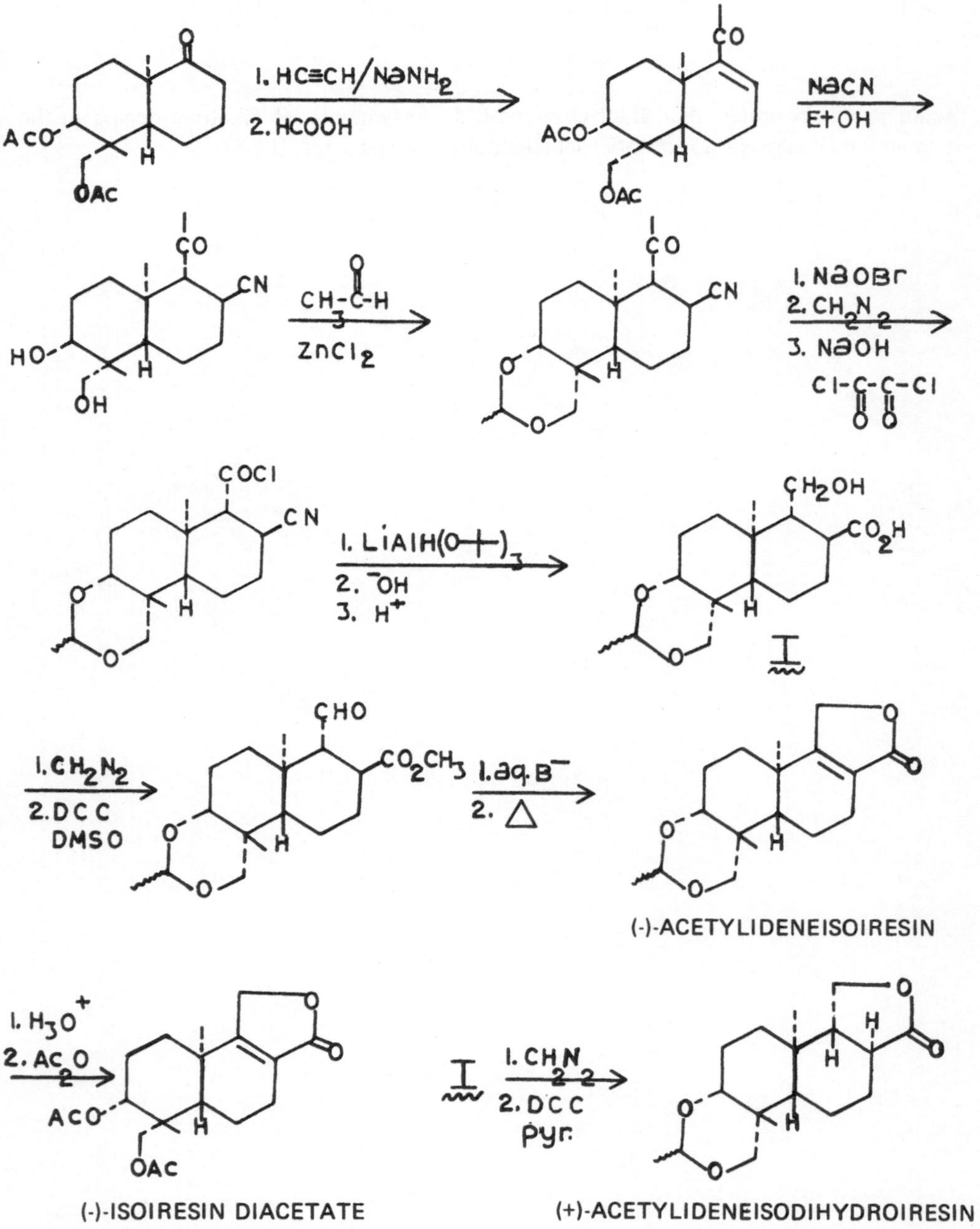

(-)-ACETYLIDENEISOIRESIN

(-)-ISOIRESIN DIACETATE

(+)-ACETYLIDENEISODIHYDROIRESIN

IIB-12   The starting material for this synthesis was previously encountered in Synthesis IIB-3. A Schuster-Meyer rearrangement of a propargyl alcohol leads to an enone which gives a 1,4-addition reaction with excess sodium cyanide in ethanol. This enables control over the stereochemistry at C-8 and C-9 (acetyl and cyano equatorial) necessary for the construction of dihydroiresin. The problem of introducing the double bond between C-7 and C-8 necessary for the synthesis of iresin remains unsolved.

IIB-13 JUNIPER CAMPHOR AND SELIN-11-EN-4α-OL

G. L. Chetty, V. B. Zalkow, and L. H. Zalkow; *Tetrahedron Letters*, 3223 (1968).

β-EUDESMOL

SELIN-11-EN-4α-OL   JUNIPER CAMPHOR

IIB-13   By starting with β-eudesmol of known absolute configuration (as shown) the absolute configuration of the optically active natural juniper camphor was demonstrated to be enantiomeric with the structure shown.

Epoxidation *trans* to the angular methyl was the crucial step in developing the proper stereochemistry at C-4.

IIB-14 NOOTKATONE

M. Pesaro, G. Bozzato, and P. Schudel; *Chem. Communs.,* 1152 (1968).

NOOTKATONE

IIB-14   The starting material is prepared from a Diels-Alder reaction of 2-ethoxy-butadiene and methyl vinyl ketone. The enol ether functionality is maintained long enough to allow for a Wittig reaction which might have been nonspecific if conducted on 4-acetylcyclohexanone.

The configuration at centers C-4 and C-7 in nootkatone is such as to dispose the methyl and isopropenyl groups equatorially. In this respect its synthesis would appear to be easier than that of eremophilone where the isopropenyl is axially disposed. The relative *trans* stereochemistry of the C-5 methyl and the isopropenyl is arranged when 2-formyl-4-isopropenylcyclohexanone is methylated. The major product (ca. 5:1) arises from apparent axial methylation (cf. IIC-8) on an enolate in which the isopropenyl is equatorially disposed.

An interesting annelation is conducted on the formyl ketone [cf. A. L. Wilds and C. Djerassi, *J. Am. Chem. Soc.*, 68, 1715 (1946)] with methylacetoacetate. Axial conjugate addition of lithium dimethylcopper (cf. I-28) leads to the wrong stereochemistry at C-4. This requires a detour involving insertion of a 3-4 double bond which is reduced by sodium borohydride in pyridine with axial hydride entry. This method [cf. S. B. Kadin, *J. Org. Chem.*, 31, 620 (1966)] appears to be an excellent one for reduction of conjugated double bonds.

IIB-15 VALERANONE

J. A. Marshall, G.L. Bundy, and W.I. Fanta; *J. Org. Chem.*, **33**, 3913 (1968).

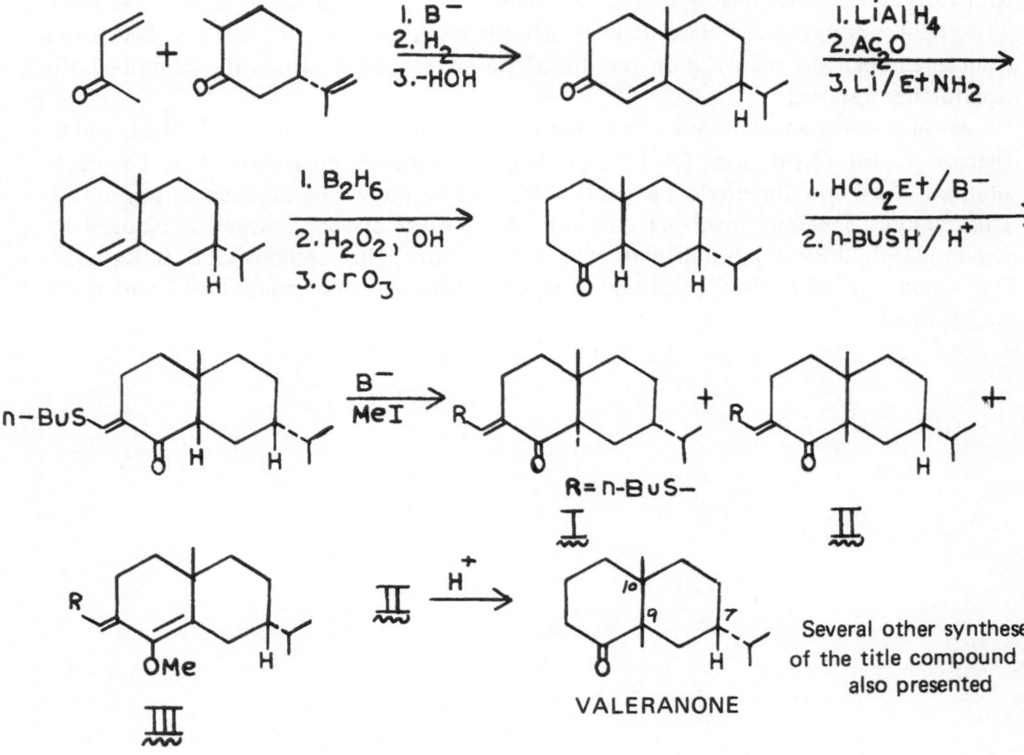

VALERANONE

Several other syntheses
of the title compound are
also presented

IIB-15   The distinctive structural feature of valeranone is the presence of two angular methyl groups in a *cis* fused decalin. The relationship of the isopropenyl at the future C-7 relative to the methyl at C-10 is arranged early via Robinson annelation of dihydrocarvone. The major product (85%) is formed through axial entry of methyl vinyl ketone which establishes a *trans* relationship between the equatorially disposed methyl and isopropenyl groups. The octalone is formed after a conformational flip (cf. IIB-9) which leads to an axial disposition of the groups. This is unlike the situation in IIB-8 where epimerization of an angular hydrogen leads to a *cis* junction-isopropenyl relationship.

The α-decalone, which is produced after appropriate manipulation, is converted to its *n*-butylthiomethylene derivative [cf. R. E. Ireland and J. A. Marshall, *J. Org. Chem.*, 27, 1615 (1962)] for purposes of encouraging angular alkylation. This is in fact realized, though only in poor yield, owing to competitive O-alkylation. To the extent that angular alkylation occurs, it does so with predominant (92:8) β attack presumably because of hindrance of the α side by the axial isopropenyl group.

The reader is referred to a recent stereospecific solution to the problem of the arrangement of the three asymmetric centers [D. J. Dawson and R. E. Ireland, *Tetrahedron Letters*, 1899 (1968)] as well as to a simpler total synthesis [E. Wenkert and D. Bergess, *J. Am. Chem. Soc.*, 89, 2507 (1967)] .

IIB-16 β-VETIVONE

J. A. Marshall and P. C. Johnson; *Chem. Communs.*, 391 (1968).

β-VETIVONE

IIB-16   A clever choice of starting material [P. J. Kropp, *J. Am. Chem. Soc.*, 87, 3914 (1965)] defines a usable stereochemical relationship of the methyl group relative to the spiro ring system. The *n*-butylthiomethylene group [cf. previous synthesis] is used to introduce an unsaturated aldehyde (cf. I-28 and IID-1 for more familiar analogs of this reaction using an enol ether of a β-diketone).

IIB-17 NOR-KETOAGAROFURAN

C. H. Heathcock and T. R. Kelly; *Chem. Communs.*, 267 (1968).

NOR-KETOAGAROFURAN

IIB-17   The starting material was that prepared for the β-eudesmol synthesis (IIB-10). The interesting feature of the synthesis is that lactone formation is employed to force epimerization about C-7, leading to an axial carbomethoxyl intermediate.

IIB-18 α-AGAROFURAN AND β-AGAROFURAN

J. A. Marshall and M. T. Pike; *J. Org. Chem.*, **33**, 435 (1968).

α-AGAROFURAN   β-AGAROFURAN

IIB-18   This approach to the agarofuran stereochemistry is different from that in IIB-17 in that it arranges for the axial isopropenyl group at C-7 through a Robinson annelation of dihydrocarvone (cf. IIB-15). Reaction of the 4-5 olefin with perbenzoic acid occurs on the same side as the angular methyl, undoubtedly because of hindrance on the β side from the isopropenyl group. The authors assume the intermediacy of an epoxide which is opened by the alcohol group. An alternative mechanism, involving *trans* addition of peroxyacid and internal alcohol to the double bond, clearly suggests itself.

Photosensitized (xylene) isomerization of the double bond is another important feature of the synthesis.

IIB-19 Δ¹⁽¹⁰⁾-ARISTOLENE

R. M. Coates, and J. E. Shaw; *Chem. Communs.*, 515 (1968).

Δ$^{1(10)}$-ARISTOLENE

IIB-19   The key intermediate, **I**, for this synthesis was also used in connection with the eremophilene (IIB-7) project. The condensation of the vinylogous amide with *trans* pent-3-ene-2-one is remarkable in that the Michael step involves the union of two rather hindered centers using a relatively dull vinylogous amide nucleophile. It should be mentioned that epimer **II** is the only one isolated if the condensation is conducted in the rather novel medium of benzene-acetic acid-water-sodium acetate [cf. R. M. Coates and J. E. Shaw, *Chem. Communs.*, 47 (1968)]. It is also interesting to note (i) monothio-ketalization of the octalindione at the enone carbonyl (cf. IIB-28) and (ii) successful conversion of the β-keto ester derivative of **I** to the keto alcohol via 2 moles of methyllithium.

The stereochemistry of the cyclopropane is controlled at the stage of the Δ^2-pyrazoline. The latter is subject to thermodynamic control at C-7 to give the equatorial dimethylcarbinyl bond.

IIB-20 ARISTOLONE

C. Berger, M. Franck-Neumann, and G. Ourisson; *Tetrahedron Letters*, 3451 (1968).

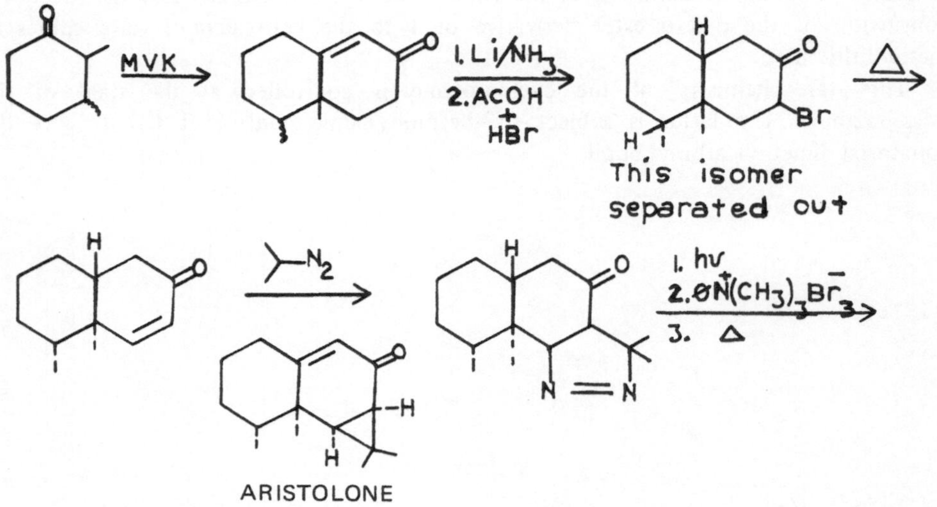

This isomer
separated out

ARISTOLONE

IIB-20   Here we encounter still another approach to the introduction of the vicinal methyl groups. Unfortunately the ratio of epimers is not given at the stage of the annelation. The *trans* ring junction of the dimethyldecalone is presumably responsible for bromination of the ketone via the Δ^2-type enol (cf. I-9 and I-29). It would have been interesting to try to form the *gem*-dimethylcyclopropane directly via a sulfonium ylid insertion [see E. J. Corey and M. Jautelat, *J. Am. Chem. Soc.*, 89, 3912 (1967)] .

IIB-21 4-DEMETHYLARISTOLONE

E. Piers, R. W. Britton, and W. De Waal; *Chem. Communs.*, 188 (1968).

4-DEMETHYLARISTOLONE

IIB-21   The key reaction employed is an insertion of a diazoketone into an intramolecular double bond [cf. G. Stork and J. Ficini, *J. Am. Chem. Soc.*, 83, 4678 (1961)]. The greater stability of the β configuration of the cyclopropane is apparently experienced at the level of the transition state for insertion.

The reader is urged to study the excellent proof of the stereochemistry of the cyclopropane ring relative to the angular methyl contained in the paper.

IIB-22 α-BOURBONENE AND β-BOURBONENE

J. D. White and D. N. Gupta; *J. Am. Chem. Soc.*, **90**, 6171 (1968).

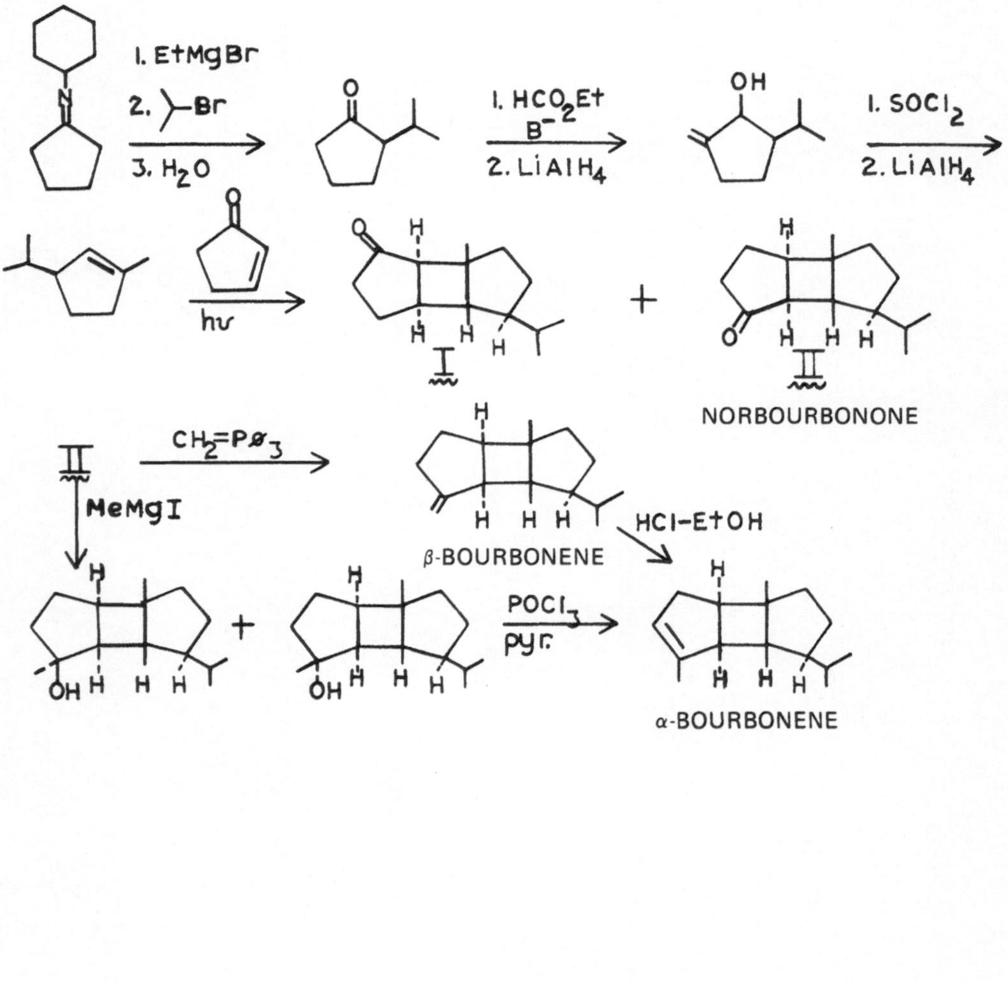

NORBOURBONONE

β-BOURBONENE

α-BOURBONENE

IIB-23 α-BOURBONENE

M. Brown; *J. Org. Chem.*, **33**, 162 (1968).

α-BOURBONENE

IIB-22, 23   These two syntheses of bourbonene illustrate different strategies. In Synthesis IIB-22 the B ring is formed from a merger of the A and C rings. Its weakest point is the orientational nonspecificity [cf. P. E. Eaton, *Accts. Chem. Res.*, 1, 50 (1968)] in the photochemical cycloaddition reaction (I:II = 52:48). Fortunately the authors devised an ingenious method for separating these two isomers based on the reluctance of the ketonic group in I to derivatize via an sp³ intermediate. The formation of β-bourbonone is then achieved via a Wittig reaction of I.

In IIB-23 the cyclobutane ring is constructed from intramolecular head-to-head photochemical cycloaddition [a head-to-tail mode would have given rise to an intermediate which could have been converted to copaene; for a synthesis of the latter see C. H. Heathcock, R. A. Badger, and J. W. Patterson, *J. Am. Chem. Soc.*, 89, 4133 (1967)]. The final ring is closed through orientationally specific internal aldolization.

Both photolyses lead to the convex stereochemistry of the isopropyl group required for bourbonene.

IIB-24 A FUNCTIONALIZED ILLUDANE

T. Matsumoto, H. Shirahama, A. Ichihara, H. Shin, S. Kagawa, N. Ito, T. Hisamitsu, T. Kamada, F. Sakan, K. Saito, S. Nishida, and S. Matsumoto; *Tetrahedron Letters*, 1925 (1968).

A functionalized illudane

IIB-24, 25   In IIB-24 a β-ketosulfoxide is employed to allow for smooth Michael addition to the cyclopentenone (cf. IIC-2). After the addition, a Pummerer rearrangement (acetic anhydride-pyridine, 1 week) leads to an acetoxy ketosulfide. The sulfur is removed under mild reductive conditions (aluminum amalgam), which allow for maintenance of the α-acetoxyketone system. The latter, amazingly, survives deketalization and aldol closure (potassium *t*-butoxide) to produce the functionalized illudane skeleton.

The high point of the synthesis of illudin M is reached by an almost bizarre transfer of an ethylene glycol unit across the molecule via thermolysis in ethanol.

This establishes the functionality required for the α-hydroxyketone system. One can well imagine the care involved in the sustenance of sensitive functionality throughout these operations. Experimental conditions will be eagerly awaited for this remarkable synthesis.

IIB-25 ILLUDIN M

T. Matsumoto, H. Shirahama, A. Ichihara, H. Shin, S. Kagawa, F. Sakan, S. Matsumoto, and S. Nishida; *J. Am. Chem. Soc.*, **90**, 3281 (1968).

ILLUDIN M

IIB-26 LINDESTRENE

H. Minato and T. Nagasaki; *J. Chem. Soc.*, (C) 621 (1968).

LINDESTRENE

IIB-26   In this synthesis as before (cf. I-9 and I-29; IIB-20) a *trans* ring junction is exploited to direct enolization (leading to enol acetate formation) in the desired (Δ^2 enol) sense. A successful Reformatsky reaction on an α-acetoxyketone should also be noted.

IIB-27 PATCHOULI AND EPIPATCHOULI ALCOHOLS

S. Danishefsky and D. Dumas; *Chem. Communs*, 1287 (1968).

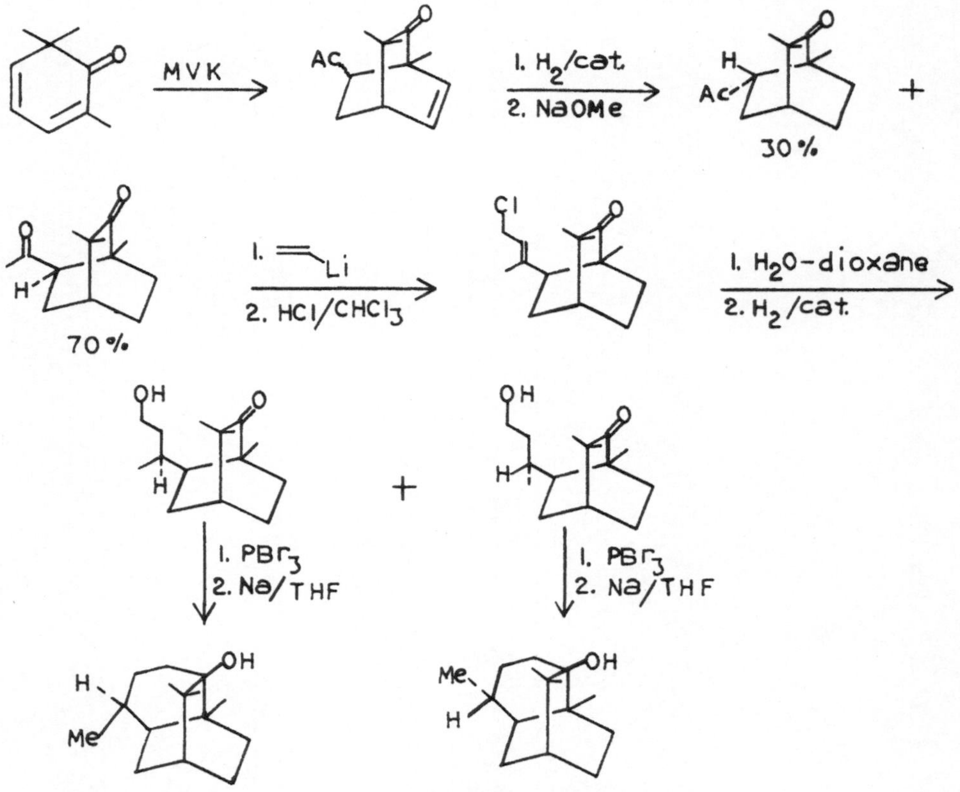
MVK
AC
1. H2/cat.
2. NaOMe
H
Ac
30%
+
70%
H
1. Li
2. HCl/CHCl3
Cl
1. H2O-dioxane
2. H2/cat.
OH
H
+
OH
H
1. PBr3
2. Na/THF
1. PBr3
2. Na/THF
OH
H
Me
Me
H
OH

PATCHOULI ALCOHOL EPIPATCHOULI ALCOHOL

IIB-27   The weakness of this synthesis (lack of control over the stereochemistry of the secondary methyl group) arose from failure of the intramolecular Grignard closure to occur at the stage of the allylic chloride. This would have allowed for convex hydrogenation leading to a β orientation of the methyl group.

IIB-28 SATIVENE

J. E. McMurry; *J. Am. Chem. Soc.*, **90**, 6821 (1968).

SATIVENE

IIB-28   Alkylation by electrophiles derived from the Wieland-Mieschner ketone is becoming a theme of increasing importance in the construction of sesquiterpenes [E.J. Corey, M. Ohno, R. B. Mitra, and P. A. Vatakencherry, *J. Am. Chem. Soc.*, 86, 478 (1964) and C.H. Heathcock, R.A. Badger, and J.W. Patterson, *ibid.*, 89, 4133 (1967)]. The synthesis of sativene is a particularly attractive application of this theme.

Aside from its architectural significance several details should be noted: (i) Low reaction temperature is helpful in promoting isopropyl Grignard addition relative to side reactions, (ii) a 2,4-dinitrophenylhydrazone is utilized as a protecting group of a ketone where sp^2 hybridization is required to avoid migration of a double bond, and (iii) convex hydroboration [cf. S. I. Sallay, *J. Am. Chem. Soc.*, 89, 6767 (1962)] is used to establish the desired stereochemistry at two asymmetric centers.

IIC-1 METHYL DEISOPROPYLDEHYDROABIETATE

T. A. Spencer, T. D. Weaver, R. M. Villarica, R. J. Friary, J. Posler, and M. A. Schwartz; *J. Org. Chem.*, **33**, 712 (1968).

METHYL DEISOPROPYLDEHYDROABIETATE

IIC-1   In contrast to the equatorial methylation in the Δ^4-4-carbomethoxy-3-octalone series (IIB-3, IIC-5, and IIIC-1), axial alkylation predominates in the 4-carbomethoxy-3-decalone system (cf. IIC-10). Clearly some rather subtle effects are operative, which are not currently understood with certainty [for an interpretation see E. Wenkert, A. Afonso, J. B. Bredenberg, C. Kaneko, and A. Tahara, *J. Am. Chem. Soc.*, 86, 2038 (1964)].

The minor product forms the basis of a new podocarpate (equatorial methyl) synthesis (see IIC-9).

IIC-2 CARNOSOL DIMETHYL ETHER AND
CARNOSIC ACID DIMETHYL ETHER

D. C. Shew and W. L. Meyer; *Tetrahedron Letters*, 2963 (1968).

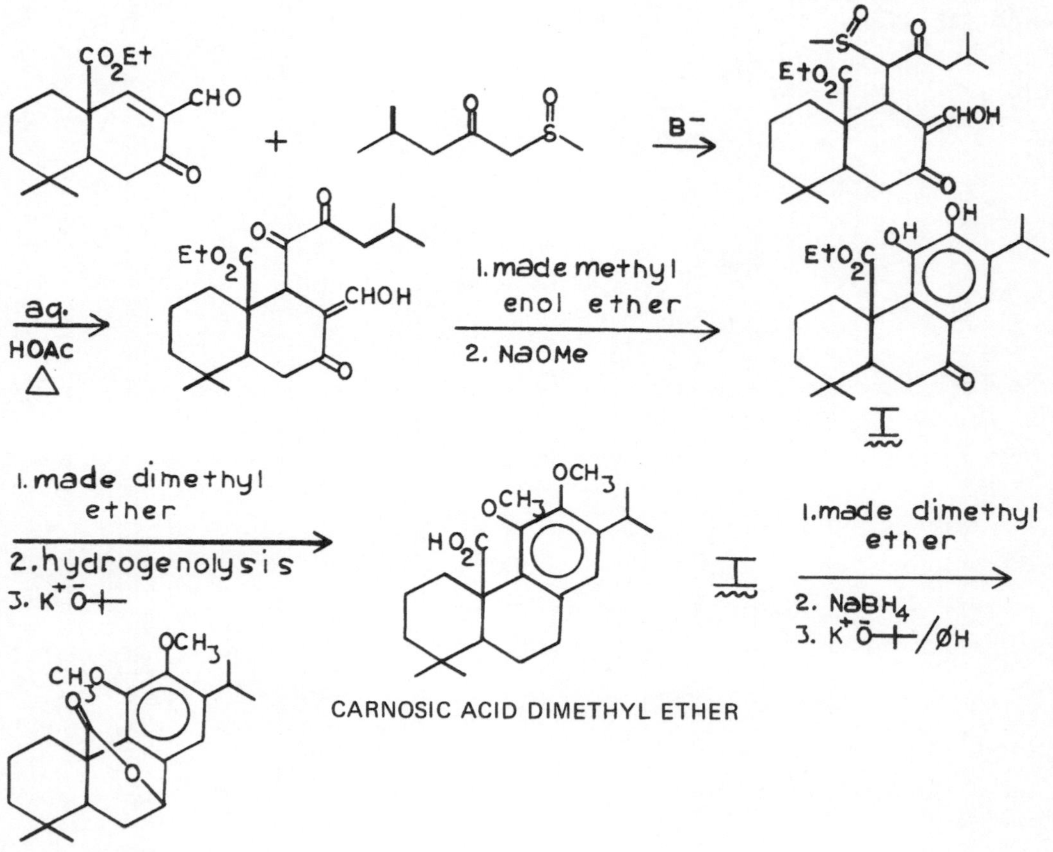

CARNOSIC ACID DIMETHYL ETHER

CARNOSOL DIMETHYL ETHER

IIC-2   Michael addition of a β-ketosulfoxide (cf. IIB-24 and IIB-25) to an alkylidene formyl ketone (formed by dehydrogenation of the formyl ketone) is the key step in producing diterpenes containing oxygenated aromatic rings. The route to α-diketones, involving Pummerer rearrangement of a β-ketosulfoxide and elimination of mercaptan from the geminal hydroxysulfide so produced, should be noted [cf. G. A. Russell and G. J. Mikol, *J. Am. Chem. Soc.*, 88, 5498 (1966)].

IIC-3 FERRUGINOL

M. Ohashi, T. Maruishi, and H. Kakisawa; *Tetrahedron Letters*, 719 (1968).

FERRUGINOL

IIC-3   The key step here is the annelation of a cyclic ketone with 4-chloromethyl-3,5-dimethylisoxazole so as to produce an α-hydroxyacetophenone system (cf. G. Stork, S. Danishefsky, and M. Ohashi, *J. Am. Chem. Soc.*, 89, 5459 (1967)] .

IIC-4   FICHTELITE

W. S. Johnson, N. P. Jensen, J. Hooz, and E. J. Leopold; *J. Am. Chem. Soc.*, **90**, 5872 (1968).

$HC\equiv C-CH_2CH_2OH$

1. NaH; ϕCH_2Br
2. EtMgBr
3. CH_2O

$\longrightarrow$ $\phi CH_2O-CH_2CH_2C\equiv C-CH_2OH$

1. PBr_3
2.

$\longrightarrow$ $\phi CH_2O-CH_2CH_2C\equiv C-CH_2CHC=O$ with CO_2Et group

1. $Ba(OH)_2$
2. $CH_2=P\phi_3$
3. Na/NH_3

$HOCH_2CH_2\underset{H}{\overset{H}{C}}=C-CH_2CH_2C=CH_2$

1. TSCl
2. LiBr

FICHTELITE

IIC-4   The key step in the synthesis of fichtelite is the polyene cyclization of a trienol to a series of closely related tricyclic intermediates. The four hydrocarbons constitute a 60% yield and the alcohol, derived from cleavage of the corresponding formate, was obtained in 34% yield. Thus tricyclic products, which belong to the same stereochemical series as C-8, C-9, and C-10, are obtained in a remarkable 94% yield. The authors interpret these results to imply the absence of a discrete, free bicyclic cationoid intermediate, i.e., the C ring is already forming as the B ring closes. They leave open the moot question of whether the process is fully concerted or whether it proceeds via bridged intermediates.

Similar methodology was employed in Synthesis IIB-9 and in Johnson's steroid synthesis (IIIF-5). The reader is urged to read two thorough reviews [W. S. Johnson, *Accts. Chem. Res.*, 1, 1 (1968) and E. E. van Tamelen, *ibid.*, 1, 111 (1968)], which set forth the results of polyene cyclization from different methods of initiation.

IIC-5 KAUR-16-EN-19-OIC ACID

K. Mori and M. Matsui; *Tetrahedron*, **24**, 3095 (1968).

TITLE COMPOUND

IIC-5   The opening steps of this synthesis, i.e., use of the Nazarov reagent and equatorial alkylation of the α-carbomethoxy-α,β-unsaturated enone, are similar to those encountered in IIB-3 (cf. IIC-1). The use of an *n*-butylthiomethylene blocking group to encourage angular alkylation with production of a *cis* fused decalone was encountered in the valeranone synthesis (IIB-15). A rather similar alkylation was also employed by Pelletier [see S. W. Pelletier, D. T. C. Chang, and A. Ogiso, *Chem. Communs.*, 830 (1968)].

IIC-6 METHYL-2-HYDROXY-1,1,4a-TRIMETHYL-
7-OXOPERHYDROPHENANTHRENE-8a-CARBOXYLATE

K. K. Mahalanabis, S. K. Mukhopadhyay, and P. C. Dutta; *Chem. Communs.*, 1641 (1968).

TITLE COMPOUND

IIC-6   Direct annelation of α,β-unsaturated ketones of the octalone type with vinyl ketones is generally considered to be a poor reaction [for a new solution, which circumvents this problem, see S. Danishefsky and R. Cavanaugh, *J. Am. Chem. Soc.*, 90, 520 (1968)]. Whether its successful realization in this case is due to the use of the β-haloketone precursor or due to special features of the intermediate enone is not clear.

IIC-7 10-CYANO-12-HYDROXY-7-OXO-17-NORPODOCARPA-
5,8,11,13-TETRAENE

W. L. Meyer, R. W. Huffman, and P. G. Schroeder; *Tetrahedron,* **24,** 5959 (1968).

TITLE COMPOUND

IIC-7   In this synthesis, Meyer utilizes Michael addition to an alkylidene formyl ketone to establish the C ring of a diterpenoid system (cf. IIC-2). The use of *t*-butylacetoacetate to avoid an undesired intramolecular Michael reaction (which forms a four-membered ring!) should be noted.

IIC-8 DESOXYPODOCARPIC ACID AND 13-METHOXYPODOCARPIC ACID

F. Giarrusso and R. E. Ireland; *J. Org. Chem.*, **33**,
3560 (1968).

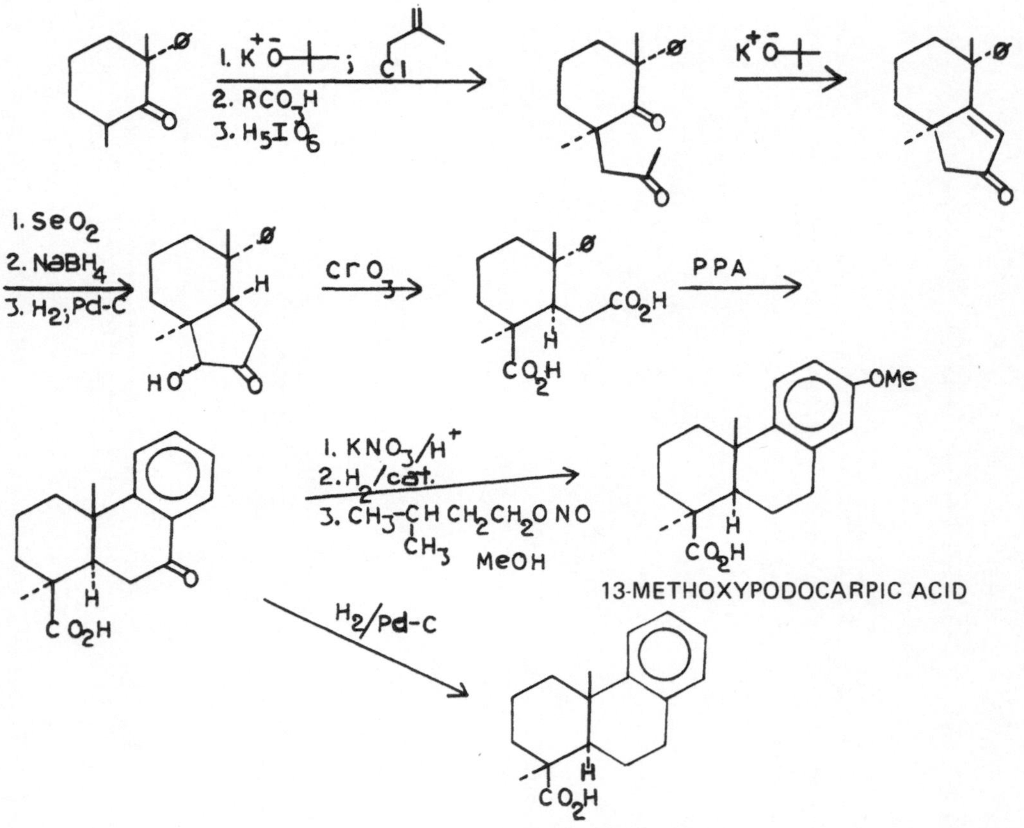

13-METHOXYPODOCARPIC ACID

DESOXYPODOCARPIC ACID

IIC-8   This approach to the podocarpic stereochemistry at C-4 (i.e., methyl at C-4 *trans* to angular methyl) should be read in relation to a related synthesis [R. E. Ireland and R. C. Kierstead, *J. Org. Chem.*, 31, 2543 (1966)] of the abietate stereochemistry (i.e., methyl at C-4 *cis* to angular methyl). In this case control at C-4 is exercized by a presumed axial alkylation of the starting enolate with methallyl chloride (cf. IIB-14). Because the phenyl group is disposed equatorially, the necessary *cis* relationship between the methyls is established. Control over the stereochemistry at C-5 is exercised when the α-hydroxyenone is reduced catalytically (cf. I-46 which gave a *trans* junction) to the *cis* fused hydroxyhydrindanone.

IIC-9 PODOCARPIC ACID

T. Spencer, R. J. Friary, W. W. Schmiegel, J. F. Simeone, and D. S. Watt; *J. Org. Chem.*, **33**, 719 (1968).

METHYL PODOCARPATE   PODOCARPIC ACID

IIC-9   The starting material for this synthesis is the minor one resulting from equatorial alkylation of the saturated β-keto ester (cf. IIC-1 and references therein). The most satisfying aspect of this synthesis is the provision for construction of the proper functionality in the aromatic ring. For an alternate approach to this functionality, which, however, avoids the serious stereochemical issues at C-4, see IIC-7. The major product (axial methyl) of the first alkylation is utilized for abietate synthesis (see IIC-1.

IIC-10   SANDARACOPIMARIC ACID

A. Afonso; *J. Am. Chem. Soc.*, **90**, 7375 (1968).

SANDARACOPIMARIC ACID

IIC-10   The key steps in this clever conversion of a steroid to a diterpene are (i) axial methylation of the β-keto ester (cf. IIC-1 and IIC-9) and (ii) dehydrohalogenation of the 16-bromo-17-keto system to afford the Δ^{14} compound which is apparently more stable than its conjugated isomer. The latter key step allows for degradation of the D ring in a most interesting fashion to give the desired axial-methyl, equatorial-vinyl stereochemistry in the C ring of the diterpene. The interesting route employed to deformylate the β-ketoaldehyde in the presence of the base labile β-diketone system should be carefully studied.

IIC-11 METHYL TRACHYLOBANATE

W. Herz, R. Mirrington, and H. Young; *Tetrahedron Letters*, 405 (1968).

METHYL
TRACHYLOBANATE

IIC-11   The orientation of the Diels-Alder reaction appears to originate by formation of the first bond [cf. R. B. Woodward and T. J. Katz, *Tetrahedron*, 5, 70 (1959)] between C-12 of the levopimarate and C-3 of the crotonate. While steric hindrance may be the decisive factor, it is also interesting to note that this mode involves cationic (or radical) character at two tertiary carbons, whereas initial attachment to C-8 would have led to such character at two secondary carbons. Maximum accumulation of double bonds would appear to be responsible for the *endo* relationship between the carbonyl group and the olefinic bridge.

The reductive cleavage of the α-acyloxyketone (cf. IIB-9 and IIIC-3) as well as the reductive π cyclization of the olefinic mesylate should be noted.

IID-1 β-AMYRIN

D.H.R. Barton, E.F. Lier, and J.F. McGhie; *J. Chem. Soc.*, (*C*) 1031 (1968).

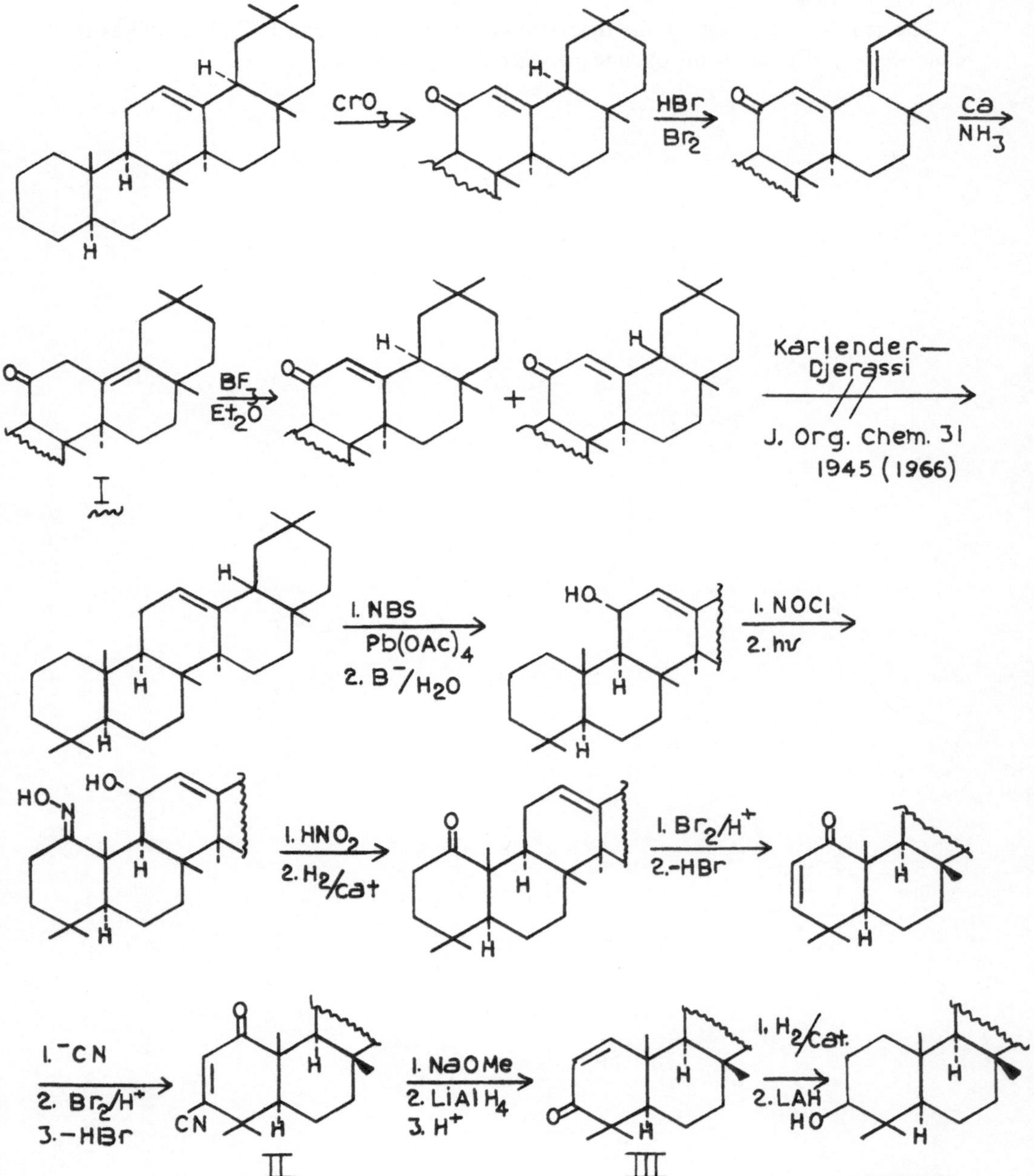

Karlender—
Djerassi

J. Org. Chem. 31
1945 (1966)

β-AMYRIN

IID-1   The starting material had previously been synthesized by several routes (see references in original paper). The key steps in its conversion to β-amyrin are (i) epimerization at C-18 via an extended enol derived from intermediate **I**, (ii) functionalization of C-1 via an 11-α-nitrite, (iii) conversion of a β-cyanoenone to a β-methoxyenone, and (iv) generation of an enone from a β-methoxyenone where the carbonyl group is placed at the original methoxy-bearing carbon (cf. I-28 and IIB-16). An alternative approach to the conversion of **II** to **III** involved (i) transformation of the nitrile function to an amide, (ii) Curtius-like degradation of the amide to an ene-urethan, and (iii) cleavage of the latter to a β-aminoenone.

IID-2 CYCLOARTENOL

D. H. R. Barton, D. Kumari, and P. Welzel; *Chem. Communs.*, 643 (1968).

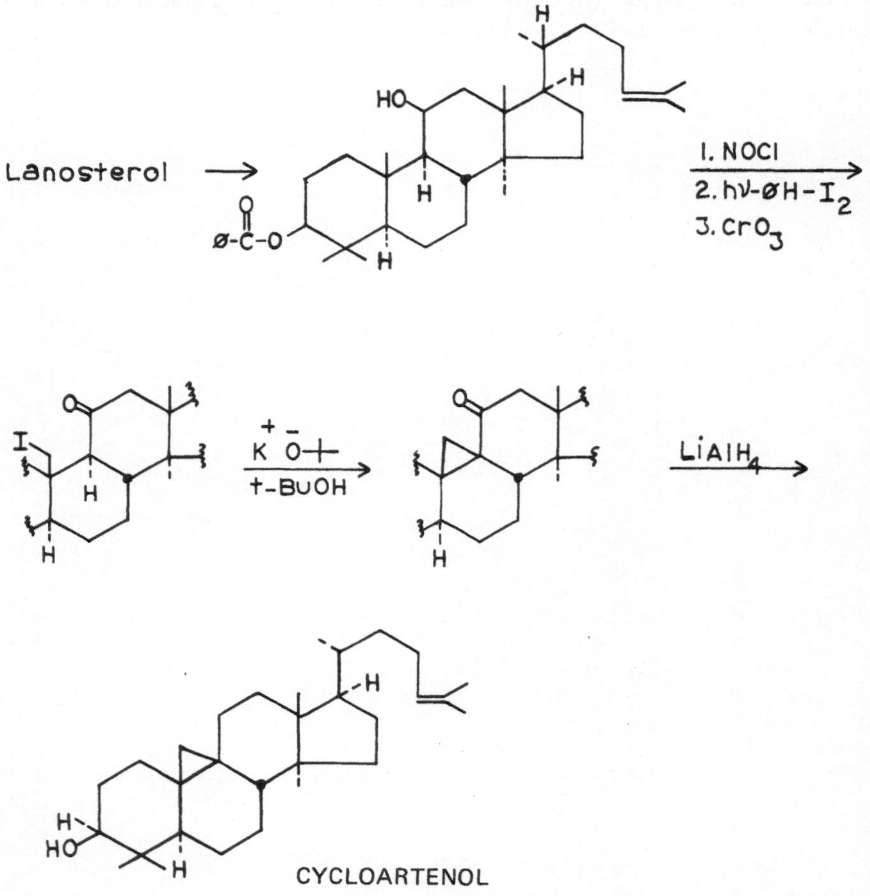

Lanosterol →

HO

ø-C-O-

1. NOCl
2. hν-øH-I$_2$
3. CrO$_3$

I

$\overset{+}{K}\ \overset{-}{O}$
t-BuOH

LiAlH$_4$

HO

CYCLOARTENOL

IID-2   The starting material has been prepared from lanosterol which had in turn been synthesized from cholesterol (see references in original paper). Inasmuch as cholesterol has been reached by total synthesis [R. B. Woodward, F. Sondheimer, D. Taub, K. Heusler, and W. M. McLamore, *J. Am. Chem. Soc.*, 74, 4225 (1952)], this work constitutes a total synthesis of cycloartenol. It will be noted that in this case, an 11β-nitrite is used to functionalize the C-19 methyl group (cf. IID-1 where an 11α-nitrite functionalized the C-1 methylene). Intramolecular alkylation (cf. IIB-28) is used to close the critical cyclopropane ring. The cyclopropyl ketone is reduced to the methylene derivative with lithium aluminum hydride (cf. benzyl ketones).

CHAPTER THREE

Other
Natural Products

IIIA-1 ANTHRAMYCIN

W. Leimgruber, A. D. Batcho, and R. C. Czajkowski; *J. Am. Chem. Soc.*, **90**, 5641 (1968).

ANTHRAMYCIN
METHYL ETHER

IIIA-1   One of the most remarkable features of this work is the survival of the
α-asymmetric center of L-hydroxyproline throughout the course of the synthesis even
though many opportunities for its racemization appear to have been available. Several
other noteworthy elements include the double bond migration (to give rise to an
enamide) which accompanies the Emmons reaction, the reduction of an ester to an
aldehyde (with diisobutylaluminum hydride), and the use of a benzoxazoline blocking
group to simultaneously protect a phenolic oxygen and an amidic nitrogen.

IIIA-2 CYCLOPENIN

H. Smith, P. Wegfahrt, and H. Rapoport; *J. Am. Chem. Soc.*, **90**, 1668 (1968).

CYCLOPENIN

IIIA-2   The crucial feature of this synthesis is the intramolecular alkylation of an amidic nitrogen (via its anion) with a proximate epoxide.

IIIA-3 6-DEMETHYL-6-DEOXYTETRACYCLINE

J. J. Korst, J. D. Johnston, K. Butler, E. J. Bianco, L. H. Conover, and R. B. Woodward; *J. Am. Chem. Soc.*, **90**, 439 (1968).

6-DEMETHYL-6-DEOXYTETRACYCLINE

IIIA-3   Several features of this remarkable synthesis should be studied carefully. An extraneous chlorine function is introduced (and subsequently removed by hydrogenolysis) to direct the intramolecular Friedel-Crafts acylation *ortho* to the methoxyl. A six-membered α-dicarbonyl ring is constructed by interpolating an oxalate unit between the 2 and 5 carbons of a 6-oxohexanoate system. The use of magnesium methoxide as a basic catalyst allows for aldol attachment of *n*-butyl glyoxylate to the 5 position of a 1,3,4-tricarbonyl system. Reductive removal of the lactonic oxygen situated α to a ketone (or γ to an α,β-unsaturated ketone) is a specific instance of a commonly used Woodward technique [see R. B. Woodward, F. E. Bader, H. Bickel, A. J. Frey, and R. W. Kierstead, *Tetrahedron*, 2, 1 (1958) and R. B. Woodward, F. Sondheimer, D. Taub, K. Heusler, and W. M. McLamore, *J. Am. Chem. Soc.*, 74, 4223 (1952)].

IIIA-4 TERRAMYCIN

H. Muxfeldt, G. Hardtmann, F. Kathawala, E. Vedejs, and J. B. Moobery; *J. Am. Chem. Soc.*, **90**, 6534 (1968).

BuLi/K$^+$Ō-t-Bu

MeO$_2$C⌣C(O)⌣CONH$_2$
(Li salt)

1. AcOH
2. base + O$_2$
3. HCl
MeOH

CH$_2$OCH$_3$

1. MeI
2. H$_3$O$^+$
3. Me$_2$SO$_4$
B$^-$

CONH$_2$

TERRAMYCIN

IIIA-4   This synthesis constitutes an advance of impressive proportions in the quest for structural and stereochemical mastery over complex organic molecules. Terramycin contains six asymmetric carbons (neglecting centers which may be involved in tautomeric keto-enol equilibria). Of these, five are contiguous and range over the A, B, and C rings. The isolated center at C-12a is dealt with last and more directly by oxygenation of the enolate from the convex side of the *cis* fused B-C system.

The *cis* relationship between the C-6 methyl and the C-5 H is established early through convex addition of methyl Grignard to the *cis* fused B-C system (cf. IIA-4). An unwanted *cis* relationship between C-5a H and C-5 H is established through the Diels-Alder reaction. Degradation converts C-4a to an aldehyde whose piperidine enamine is equilibrated to give a convex immonium salt (i.e., C-5 H *trans* to C-5a H) which is cleaved (silica gel) to the convex aldehyde. The stereochemical situation at C-4a and C-4 is less straightforward. It will be noted that the *trans* B-E fusion necessitates an axial conformation for the hydrogen at C-5a and the bridgehead double bond at the A-B junction imposes a similar requirement on the hydrogen at C-4a. These are therefore necessarily *cis* if ring B is to be a half chair. Thus, under thermodynamic control (which is possible at various stages of the condensation through Michael-retro Michael equilibria) the observed result would be expected. Indeed it may be that the Michael adduct, which would give a *trans* relationship between C-5a H and C-4a H, does not in fact undergo the Dieckmann closure of the B ring. The configuration at C-4 is also, in principle, subject to the thermodynamic control (via the C-3 or C-1 ketone) producing the normal [see J. Donohue, J. D. Dunitz, K. N. Trueblood, and M. S. Webster, *J. Am. Chem. Soc.*, 85, 851 (1963)] tetracycline stereochemistry at this center.

Aside from the stereochemical considerations, the reader is urged to note (i) the use of the CH$_2$-O-CH$_3$ blocking group for the phenol, (ii) the use of the building block, methyl 3-oxoglutaramate, and (iii) the autoxidation of the 12a position in 32% yield (cf. 6.5% in IIIA-3 with cerous chloride-oxygen) using alkaline conditions.

The reader is above all urged to read the references shown as well as another one [H. Muxfeldt and W. Rogalski, *J. Am. Chem. Soc.*, 87, 933 (1965)] to develop the proper mood necessary to appreciate this masterpiece.

IIIA-5 DIHYDRORADICININ

K. Kato and Y. Hirata; *Chem. Communs.*, 319 (1968).

DIHYDRORADICININ

IIIA-5   The inference from the paper is that acetone dicarboxylic acid is twice butyrylated and that subsequent heating gives the anhydride. Hydrolysis-decarboxylation-recyclization and de-butyrylation gives the required γ-hydroxy-α-pyrone. Compound **I**, arising from C-acylation and O-alkylation is the minor one (11%) relative to **II** which arises from O-acylation and C-alkylation (35%). A similar issue is encountered in IIIB-3.

IIIA-6 FULVOPLUMIERIN

G. Büchi and J. A. Carlson; *J. Am. Chem. Soc.,* **90**, 5336 (1968).

FULVOPLUMIERIN

IIIA-6   The starting material was prepared by a Diels-Alder condensation of dimethyl penta-2,3-dienoate with 1,3-butadiene. The vinylogous malonic ester is condensed with the dimethylacetal of dimethylformamide. In synthetic terms this is analogous to a Vilsmeier reaction in its introduction of a formyl equivalent, though apparently more useful on sensitive substrates.

The direction of the aldolization of the dialdehyde is apparently governed by steric factors. The authors point out that the isomeric hydroxyfulvene (which would arise via condensation of the methylene group of the vinylogous β-dicarbonyl) cannot be constructed with molecular scale models. The hydroxyfulvene is converted to the chloride which is coupled (addition-elimination) to lithium di-(*trans*-1-propenyl)cuprate.

IIIA-7 GEOSMIN

J. A. Marshall and A. R. Hochstetler; *J. Org. Chem.,* **33**, 2593 (1968).

GEOSMIN

IIIA-7   It is interesting to note that both Δ^4-10β-methyloctalin and Δ^4-6α,10β-dimethyloctalin give the same (57:43) ratio of 4α,5α:4β,5β-oxides.

IIIA-8 INDOLMYCIN

M. Preobrazhenskaya, E. Balashova, K. Turchin, E. Padeiskaya, N. Uvarova, G. Pershin, and N. Suvorov; *Tetrahedron*, **24**, 6131 (1968).

INDOLMYCIN

IIIA-8   The displacement of the isopropylamino group with malonate anion must undoubtedly involve participation from the indolic nitrogen (cf. I-6 and IIID-3).

IIIA-9 KASUGAMYCIN

Y. Suhara, F. Sasaki, K. Maeda, H. Umezawa, and M. Ohno; *J. Am. Chem. Soc.,* **90**, 6559 (1968).

KASUGAMYCIN

IIIA-9   The   addition   of   nitrosyl   chloride   to   *trans*-5-acetimido-6-methyl-5,6
-dihydropyran appears to occur with exclusive axial entry of the nitrosyl group. The *cis*
relationship of the anomeric chlorine and the nitrosyl group presumably reflects
thermodynamic control in that it allows for three of the four substituents to be
equatorially disposed. Inversion of the anomeric carbon by D-inositol gives the correct
relative stereochemistry in the A ring and results in its kinetic resolution.

IIIA-10 5,6-*Trans*-PENICILLIN V METHYL ESTER

A. K. Bose, G. Spiegelman, and M. S. Mankas; *J. Am. Chem. Soc.*, **90**, 4506 (1968).

6-*EPI*-PENICILLIN V METHYL ESTER

IIIA-10   The mechanism of this interesting route to β-lactams is unclear. The most obvious possibility is formation of azidoketene by dehydrochlorination of azidoacetyl chloride which undergoes cycloaddition with the imine. However, the actual course may be more complicated [see A. K. Bose, B. Anjaneyulu, S. K. Bhattacharya, and M. S. Manhas, *Tetrahedron*, 23, 4769 (1967)].

IIIA-11 PYRETHROLONE

L. Crombie, P. Hemesley, and G. Pattenden; *Tetrahedron Letters*, 3021 (1968).

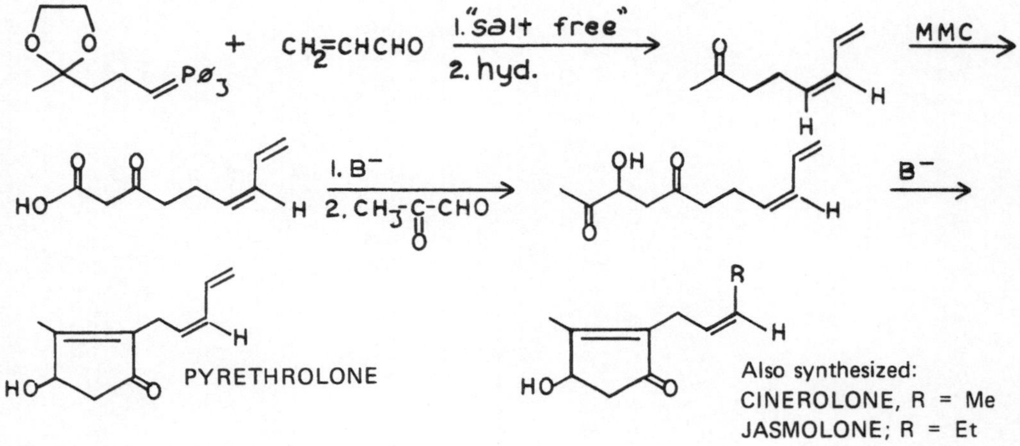

PYRETHROLONE

Also synthesized:
CINEROLONE, R = Me
JASMOLONE; R = Et

IIIA-11   The bromoketal precursor of the starting phosphorane was previously encountered (see I-26, I-27). Wittig reaction with acetaldehyde, under "salt-free" conditions [cf. M. Schlosser and K. F. Christmann, *Annalen,* 708, 1 (1967)] gave the *cis* diene. This is condensed with pyruvaldehyde after activation of the methyl group via magnesium methyl carbonate. The condensation is conducted on the potassium salt of the β-keto acid. It is interesting to note that the potentially labile diketodienol survives intramolecular aldolization.

IIIA-12 RHODOQUINONE

G. D. Daves, J. J. Wilczynski, P. Friis, and K. Folkers, *J. Am. Chem. Soc.*, **90**, 5587 (1968).

RHODOQUINONE

IIIA-12 Addition of ammonia to quinone **I** is apparently accompanied by air oxidation to give rhodoquinone.

IIIB-1 ALLOEVODIONOL, CANNABICHROMENE, AND FRANKLINONE

G. Cardillo, R. Cricchio, and L. Merlini; *Tetrahedron,* **24,** 4825 (1968).

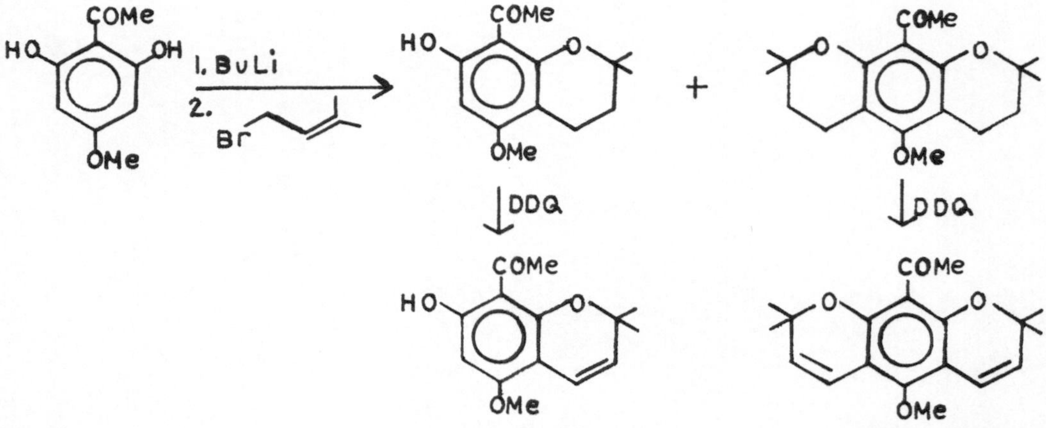

ALLOEVODIONOL

FRANKLINONE

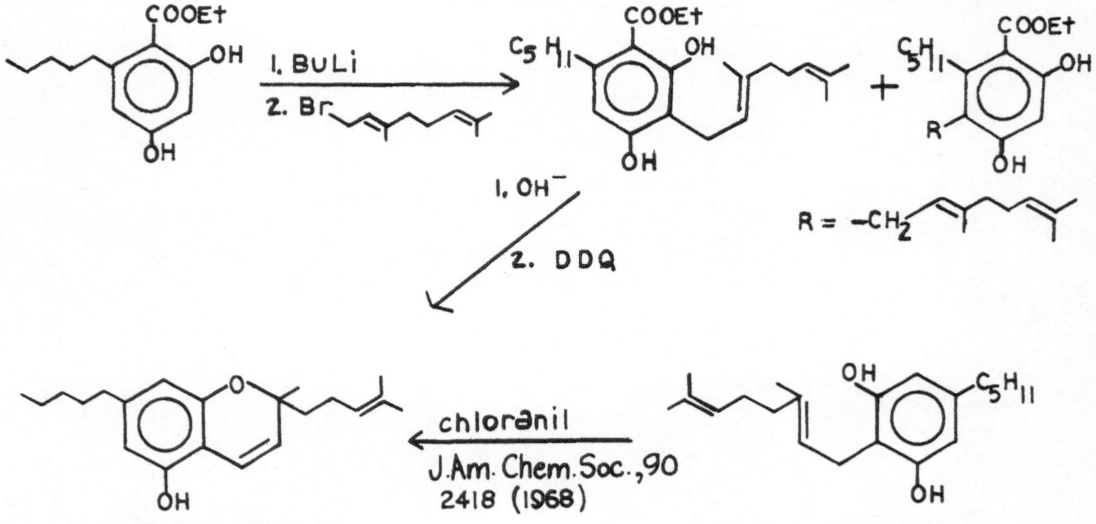

$R = -CH_2$

CANNABICHROMENE

chloranil

J.Am. Chem. Soc.,90
2418 (1968)

CANNABIGEROL

IIIB-1   The lithium salt of the quinone is utilized in a nonpolar solvent such as benzene to promote maximum (15 to 20%) *ortho*-C-alkylation. In the syntheses of alloevodionol and franklinone the c-isoprenylated phenols cyclize to chromanes on silica gel chromatography. The latter are dehydrogenated with 2,3-dichloro-5,6-dicyanobenzo-quinone (DDQ).

The synthesis of cannabichromene illustrates a different approach. Here the alkylated phenol is cyclized to the chromene by loss of hydride with DDQ. These authors invoke the intermediacy of an *o*-quinone methide which undergoes valence isomerization to the chromene.

A more searching analysis of the factors governing quinone dehydrogenation (using chloranil) of hydroaromatics in the cannabinol series, as well as a closely related synthesis of cannabichromene, was conducted by R. Mechoulam, B. Yagnitinsky, and Y. Gaoni, *J. Am. Chem. Soc.*, 90, 2418 (1968). They propose a concerted cyclization.

IIIB-2 CANNABICHROMENE AND CANNABICYCLOL

V. V. Kane and R. K. Razdan; *J. Am. Chem. Soc.,* **90**, 6551 (1968).

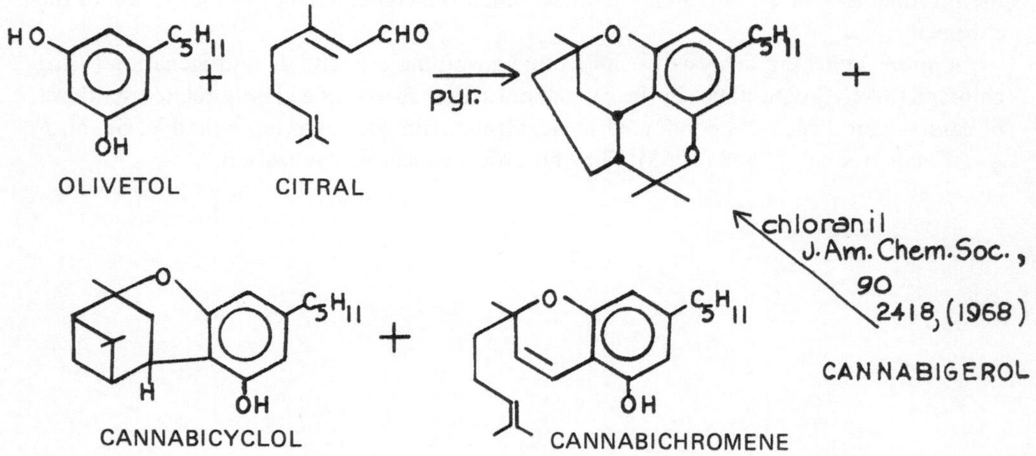

OLIVETOL   CITRAL

pyr.

chloranil
J. Am. Chem. Soc.,
90
2418, (1968)

CANNABIGEROL

CANNABICYCLOL   CANNABICHROMENE

IIIB-2   This approach to cannabichromene is essentially identical to another [L. Crombie and R. Ponsford, *Chem. Communs.*, 894 (1968)] published this year, except that cannabicyclol is formulated differently.

The formation of cannabichromene may be rationalized as arising from aldol condensation of the resorcinol derivative (at the carbon between the oxygens) with the aldehyde of citral followed by $S_N 2'$ type displacement of water by the phenolic oxygen. This seems more likely than a reverse sequence involving Michael addition from the phenolic oxygen (to the hindered disubstituted carbon) followed by aldolization. The tetracyclic diether could well arise from cannabichromene. The diether was also obtained by chloranil treatment of cannabigerol (see IIIB-1).

Cannabicyclol may arise through aldol condensation of olivetol and citral followed by dehydration. The intermediate so produced could give the product by the path shown.

IIIB-3   Δ6a,10a-TETRAHYDROCANNABINOL

U. Claussen, P. Mummenhoff, and F. Korte; *Tetrahedron*, **24**, 2897 (1968).

Δ6a,10a-TETRAHYDROCANNABINOL

IIIB-3   This paper is a more thorough investigation of a reaction first reported in 1941 [see R. Gosh, A. R. Todd, and D. C. Wright, *J. Chem. Soc.*, 137 (1941)]. In addition to the cannabinol system, formed by conjugate addition from oxygen and aldol closure from carbon (cf. IIIB-2), an isomer was isolated whose genesis involves the reverse modes of attachment of carbon and oxygen to pulegone.

IIIB-4   TETRAHYDROCANNABINOL CONGENER

H. S. Aaron and C. P. Ferguson; *J. Org. Chem.*, 33, 684
(1968).

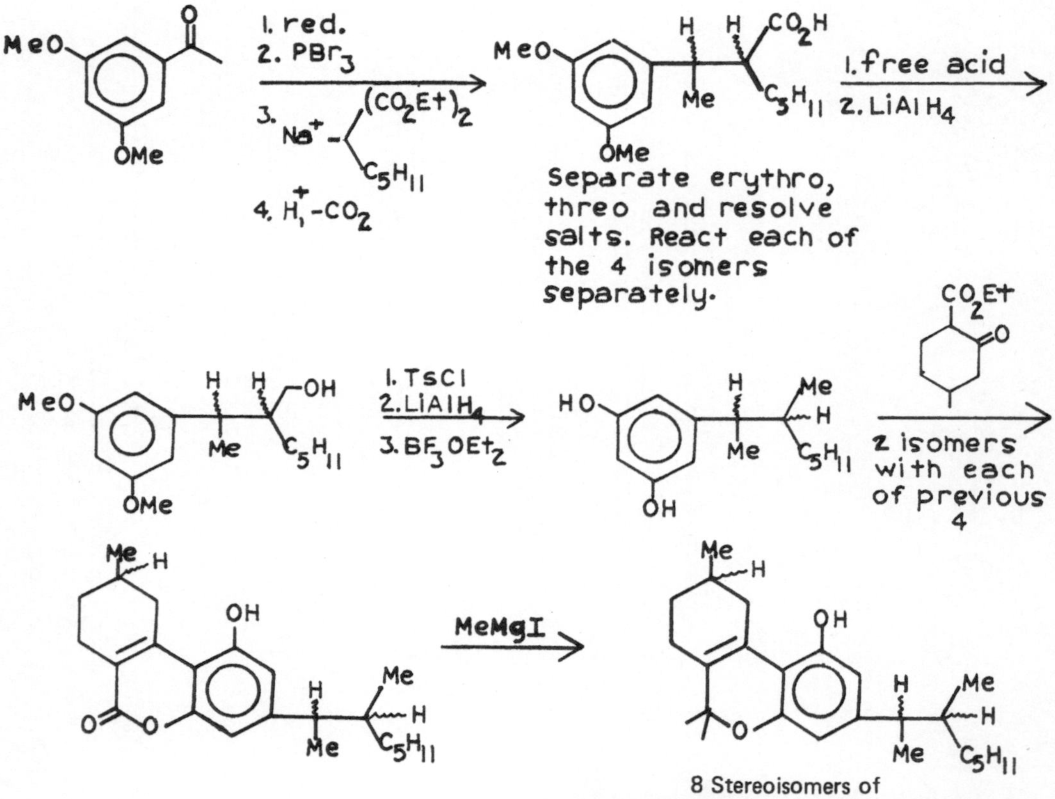

Separate erythro,
threo and resolve
salts. React each of
the 4 isomers
separately.

2 isomers
with each
of previous
4

8 Stereoisomers of
TETRAHYDROCANNABINOL CONGENER

IIIB-4   The method of attachment involves C-aldolization and O-acylation with a resorcinol derivative. The *gem*-dimethyl group is introduced via Grignard addition to a lactonic carbonyl. The second methyl group might have been introduced via an intermediary bicyclic ketophenoxide or, alternatively, through a tricyclic oxonium intermediate derived from the first formed magnesiohemiacetal.

IIIC-1 STEREOSPECIFIC APPROACH TO JUVENILE HORMONES

R. Zurfluh, E. N. Wall, J. B. Siddall, and J. A. Edwards; *J. Am. Chem. Soc.,* **90**, 6224 (1968).

trans,cis-6-ETHYL-10-
METHYLDODECA-5,9-DIEN-2-ONE

IIIC-2 CECROPIA JUVENILE HORMONE

W. S. Johnson, T. Li, D. J. Faulkner, and S. F. Campbell; *J. Am. Chem. Soc.*, **90**, 6225 (1968).

CECROPIA JUVENILE HORMONE

IIIC-3 CECROPIA JUVENILE HORMONE

E. J. Corey, J. A. Katzenellenbogen, N. W. Gilman, S. A. Roman, B. W. Erickson; *J. Am. Chem. Soc.*, **90**, 5618 (1968).

CECROPIA JUVENILE HORMONE

IIIC-1-3   A successful stereospecific synthesis of juvenile hormone must provide for (i) a *cis* relationship between the terminal ethyl function and the alkenyl chain in an epoxide ring (or, presumably, in a double bond precursor), (ii) a *cis* relationship between the ethyl group at C-7 and the alkenyl chain (i.e., a *trans* Δ^6 double bond), and (iii) a *cis* relationship between the methyl at C-3 and the terminal carbomethoxy group. It is interesting to note how these objectives were reached in Syntheses IIIC-1-3.

In Model Synthesis IIIC-1, relationships (i) and (ii) are attained through two Wharton fragmentations. Equatorial methylation (cf. IIB-3) and axial delivery of hydride (via lithium tri-*t*-butoxyaluminum hydride) establishes a *cis* relationship between the methyl at C-11 and the hydrogen at C-10. Wharton fragmentation [cf. P. S. Wharton, *J. Org. Chem.*, 26, 4781 (1961)] preserves this juxtaposition in the double bond which is formed, thereby satisfying relationship (i). The condition for relationship (ii) is satisfied at the stage of borohydride reduction of the original hydrindenedione. This establishes a *trans* relationship between the ethyl at C-7 and the hydrogen at C-6 which is transmitted to the olefin in the second Wharton opening.

Synthesis IIIC-2 secures relationship (iii) in a most straightforward fashion by employing a *trans* γ-bromocrotonate system as an alkylating agent. Condition (ii) is satisfied by a Johnson stereospecific opening of a cyclopropyl carbinyl bromide [cf. S. F. Brady, M. A. Ilton, and W. S. Johnson, *J. Am. Chem. Soc.*, 90, 2882 (1968) and Synthesis IIIF-5]. Relationship (i) is achieved via formation of a *threo* chlorohydrin upon addition of methyl Grignard to the α-chloroketone [cf. J. W. Cornforth, R. H. Cornforth, and K. K. Mathew, *J. Chem. Soc.*, 112 (1959) and D. J. Cram and F. A. A. Elhafez, *J. Am. Chem. Soc.*, 74, 5828 (1952)].

In Synthesis IIIC-3 selective ozonization of a substituted methoxy cyclohexadiene is employed to unambiguously afford a *cis* relationship between the ethyl at C-11 and C-9 [relationship (i)]. The *trans* 6,7 and 2,3 double bonds are each generated from acetylenic carbinol precursors by a stereospecific sequence involving (a) reduction with lithium aluminum hydride and iodination to give a *trans* iodoalkene [see E. J. Corey, J. A. Katzenellenbogen, and G. H. Posner, *J. Am. Chem. Soc.*, 89, 4245 (1967)] and stereospecific coupling of the iodoalkene with a lithium dialkylcuprate [see E. J. Corey and G. H. Posner, *ibid.*, 90, 5615 (1968)]. An elegant, mild, conversion of the allylic alcohol to the α,β-unsaturated ester [see E. J. Corey, N. W. Gilman, and B. E. Ganem, *ibid.*, 90, 5616 (1968)] should also be noted. Terminal chlorohydrin formation is achieved utilizing N-bromosuccinimide in aqueous monoglyme, taking advantage of hydrophobic bonding which exposes the terminal double bond to attack while shielding the interior regions [cf. E. E. van Tamelen and T. J. Curphey, *Tetrahedron Letters*, 121 (1962)].

IIIC-4 PROPYLURE

G. Pattenden; *J. Chem. Soc.*, (*C*) 2385 (1968).

PROPYLURE

IIIC-4   The key stereochemical feature of this synthesis is the use of an acetylenic group as the progenitor (after Na/NH$_3$ reduction) of the *trans* disubstituted double bond. The use of β-bromopropionaldehyde dimethylacetal as a three-carbon alkylating agent should be noted.

IIID-1 DEOXOPHYLLOERYTHROETIOPORPHYRIN

M. E. Flaugh and H. Rapoport; *J. Am. Chem. Soc.*, **90**, 6877 (1968).

DEOXOPHYLLOERYTHROETIOPORPHYRIN

$$\underset{\sim}{III}$$

a= R=$CO_2CH_2C_6H_4OCH_3$
R'=CH=NH·HCl

b = R=$CO_2CH_2C_6H_4OCH_3$
R'=CHO

c = R=$CO_2CH_2C_6H_4OCH_3$
R'=CHS

IIID-1   Several features of this synthesis should be noted. The use of a malononitrile blocking group to immobilize a pyrrole aldehyde toward condensation has precedent [see R. B. Woodward, *Pure and Applied Chem.*, 2, 383 (1961)]. Control over the mode of condensation of intermediate **I** was assured by the presence of only one unsubstituted α-pyrrole position. However, it is most interesting that the cyclopentanone carbonyl group is a more receptive electrophile than the aldimonium group in intermediate **II**.

In the final closure, the thioaldehyde (III$_C$) was not noticeably superior to the aldehyde (III$_B$) as an electrophile (cf. reference cited above) at least in terms of yield of product produced from the final cyclization.

IIID-2 OCTAETHYLPORPHYRIN

H. W. Whitlock and R. Hanauer; *J. Org. Chem.,* **33**, 2169 (1968).

OCTAETHYLPORPHYRIN

IIID-2   Both substitution and oxidation are necessary for the success of this cyclotetramerization reaction. Thus omission of oxygen did not lead to isolation of the corresponding octaethyl tetrapyrrole nor could porphyrin itself be detected using the same sequence on 2-dimethylaminopyrrole.

IIID-3 SYNTHESIS OF OXYPORPHYRINS

A.H. Jackson, G.W. Kenner, G. McGillivray, and K.M. Smith; *J. Chem. Soc.*, (C) 294 (1968).

AN OXYPORPHYRIN

IIID-3   Intermediates **I** and **II** are produced from the same type of alkylation-decarboxylation as is involved in the first step of Synthesis IIID-1. Pyridine is an effective leaving group for triggering the condensation. The formulation, shown below, is one of several related possible pathways.

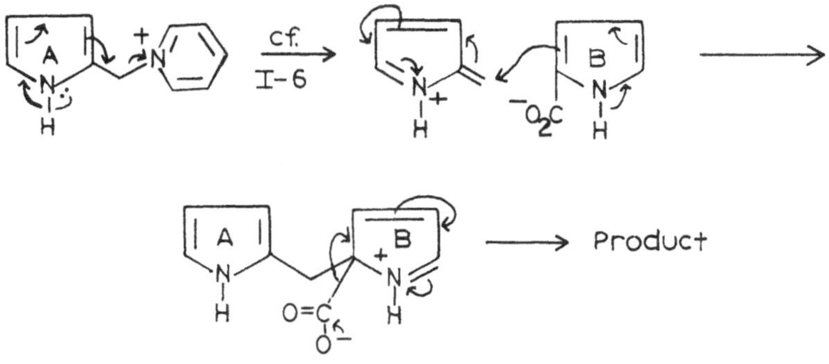

IIIE-1 PROSTAGLANDINS E_1, A_1, $F_{1\alpha}$, AND $F_{1\beta}$

E. J. Corey, N. H. Andersen, R. M. Carlson, J. Paust, E. Vedejs, I. Vlattas, and R. E. K. Winter; *J. Am. Chem. Soc.*, **90**, 3245 (1968).

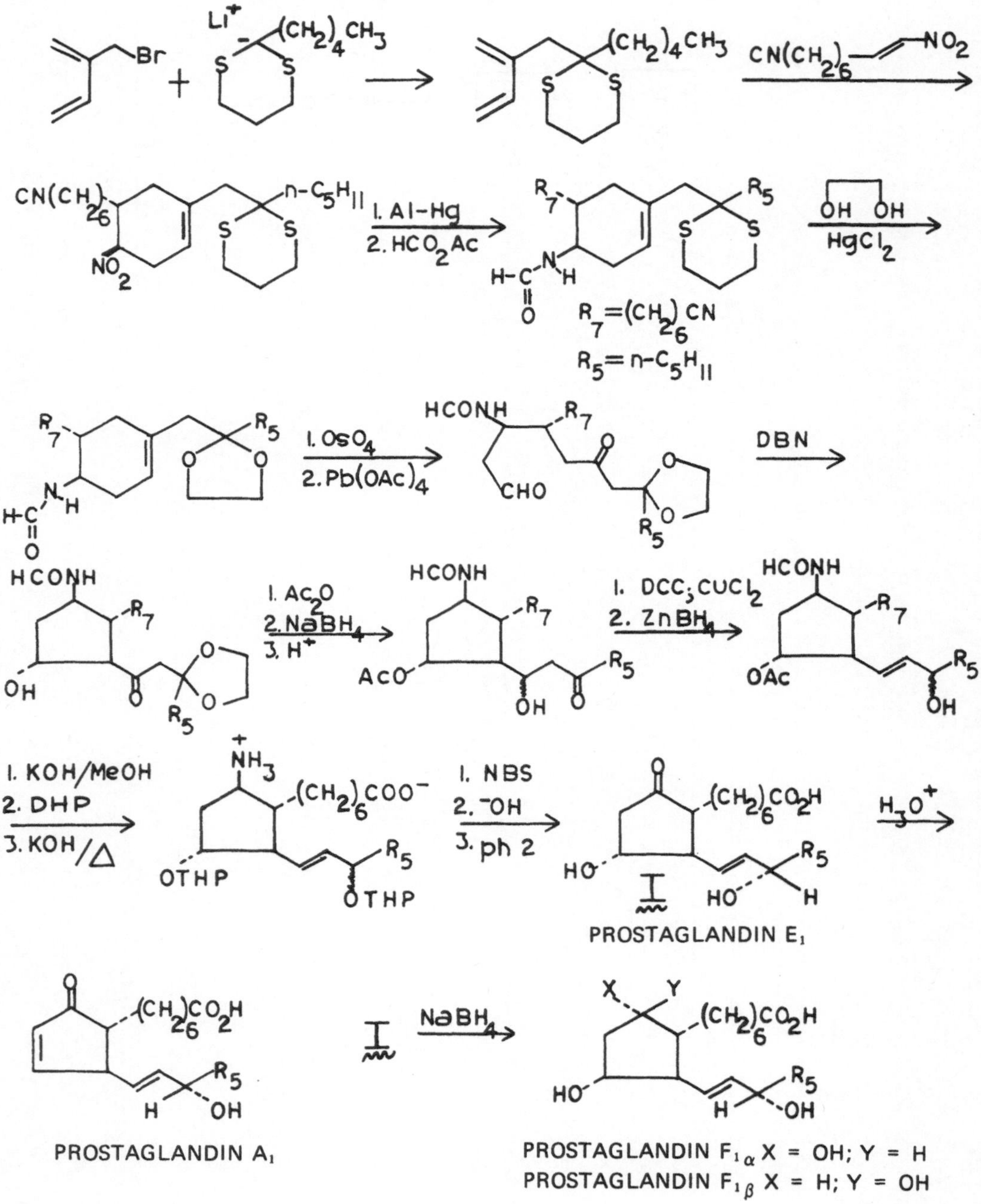

PROSTAGLANDIN E₁

PROSTAGLANDIN A₁

PROSTAGLANDIN $F_{1\alpha}$ X = OH; Y = H
PROSTAGLANDIN $F_{1\beta}$ X = H; Y = OH

IIIE-1   The key steps in Synthesis IIIE-1 are (i) the use of a dithiane derivative to generate nucleophilic capabilities at an acyl carbon [cf. E. J. Corey and D. Seebach, *Angew. Chem. Int. Ed. Engl.*, 4, 1075 (1965)], (ii) the direct replacement of dithiane blocking group by a dioxolane derivative, (iii) the use of 1,5-diazabicyclo [4,3,0]-5-nonene (DNB) as a catalyst for effecting intramolecular aldolization which is not accompanied by β elimination, (iv) the use of the dicyclohexylcarbodiimide (DCC)-cupric chloride-ether system as a mild method for effecting dehydration of a β-hydroxyaldehyde, and (v) the use of N-bromosuccinimide for effecting the conversion of an amine to a ketone. For a recent elegant method for achieving the transformation see E.J. Corey and K. Achiwa, *J. Am. Chem. Soc.* 91, 1429 (1969).

IIIE-2 PROSTAGLANDINS OF THE E_1 AND F_1 SERIES

E. J. Corey, I. Vlattas, N. H. Andersen, and K. Harding; *J. Am. Chem. Soc.*, **90**, 3247 (1968).

$R_7 = (CH_2)_6 CN$

$R_5 = (CH_2)_4 CH_3$

1. Al–Hg
2. HCOOAc

$\downarrow$ TsOH

1. $ZnBH_4$
2. DHP
3. KOH–MeOH–H_2O
4. NBS
5. B^-
6. ph 2

PROSTAGLANDIN E_1
+
15-EPIPROSTAGLANDIN E_1

11-EPIPROSTAGLANDIN E_1
+
11,15-EPIPROSTAGLANDIN E_1

PROSTAGLANDIN E_1
Also synthesized were some prostaglandins of the F_1 series

IIIE-2   The key step in this synthesis is the reaction of the bisdioxolane with tosyl acid in acetone to afford a vinylogous β-hydroxyketone. Whether this closure occurs via a dialdehyde or via an intermediate in the deketalization pathway is unclear. It is interesting to note that Michael addition using a ketal (presumably via an enol ether intermediate) was achieved in Synthesis I-26. One permutation of several related possibilities is shown below.

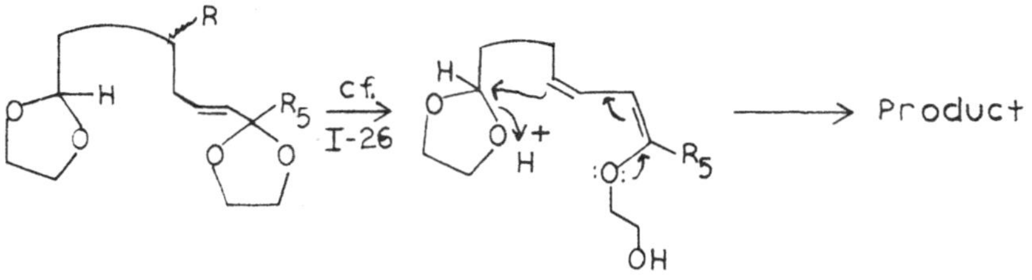

The advantage of such a formulation is that it accounts for the actual pathway relative to others which would appear to be quite competitive via a ketoaldehyde intermediate.

IIIE-3 PROSTAGLANDINS

W. P. Schneider, U. Axen, F. H. Lincoln, J. E. Pike, and J. L. Thompson; *J. Am. Chem. Soc.*, **90**, 5895 (1968).

$R_5 = (CH_2)_4 CH_3$

Synthesized in a related fashion were many other prostaglandins

IIIE-3   The stereochemistry of the three contiguous asymmetric centers on the cyclopentane ring is arranged in a most clever fashion. Predominant convex alkylation of the bicyclo[3,1,0]hexane system by iodoheptanoate establishes the required *trans* relationship between the heptanoate side chain and the cyclopropane ring. The latter is opened with inversion at the least-hindered carbon during solvolysis of the cyclopropyl-carbinyl mesylate to give the required *trans-trans* relationship of the hydroxyl C-8 and C-7 functions.

This report vindicates, at least qualitatively, a previous report [G. Just and C. Simonovitch, *Tetrahedron Letters*, 2093 (1967)] which has recently been disputed [K. G. Holden, B. Hwang, K. R. Williams, J. Weinstock, M. Harman, and J. A. Weisbach, *ibid.*, 1569 (1968)]. Apparently conditions required for opening of the cyclopropylcarbinyl system (mesylate or epoxide) are quite delicate.

IIIE-4   7-OXAPROSTAGLANDIN F$_{1\alpha}$

J. Fried, S. Heim, S. J. Etheredge, P. Sunder-Plassmann, T. S. Santhanakrishnan, J. Himizu, and C. H. Lin; *Chem. Communs.*, 634 (1968).

7-OXAPROSTAGLANDIN F$_{1\alpha}$

IIIE-4   Very clever use was made of a dialkyl alkynyl aluminum reagent (complexed to triethylamine) to effect a *trans* opening of an epoxide. This provides the required *cis-trans-trans* stereochemistry at carbons 9, 8, 12, and 11, respectively. The only step lacking stereochemical control is that in which the C-15 hydroxyl is introduced through allylic oxidation via selenium dioxide. It is interesting to note that the latter reaction can be conducted in the presence of two free hydroxyl groups.

The successful application of this approach to the synthesis of prostaglandins would require preparation of a β-oriented leaving group at C-8 which would be inverted by a carbon nucleophile. Such a sequence would not appear to be unattainable.

IIIF-1   (−)-17β-HYDROXY-Δ$^{9(10)}$ DES A-ANDROSTEN-5-ONE

Z. G. Hajos, D. R. Parrish, and E. P. Oliveto; *Tetrahedron*, **24**, 2039 (1968).

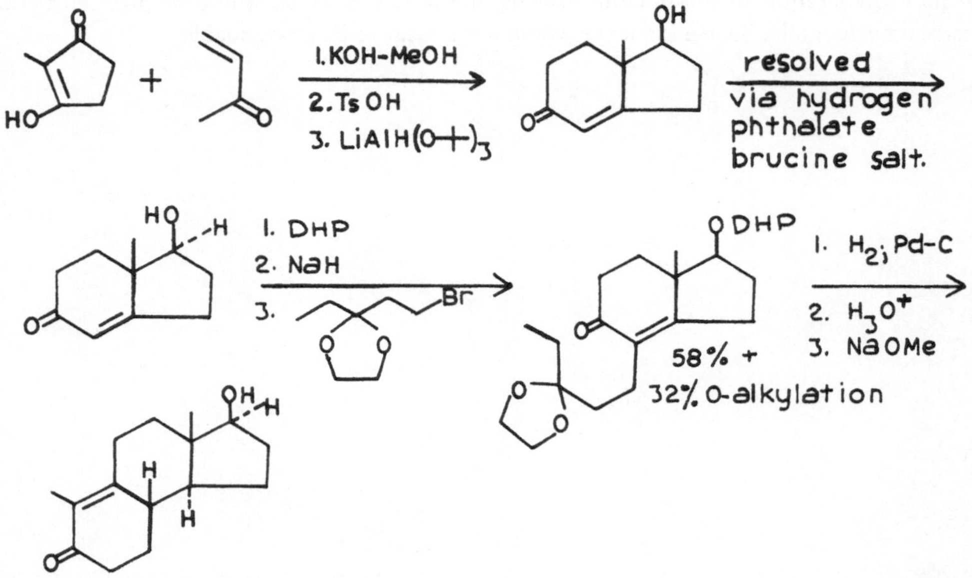

1.KOH–MeOH
2.Ts OH
3. LiAlH(O┼)₃

resolved
via hydrogen
phthalate
brucine salt.

1. DHP
2. NaH
3. ⟨Br structure⟩

1. H₂; Pd–C
2. H₃O⁺
3. NaOMe

58% +
32% O-alkylation

TITLE COMPOUND

IIIF-1   The key step in this synthesis is catalytic reduction of the Δ^4-hydrindenolone bearing a large substituent at C-4 and a β-oriented tetrahydropyranyl ether at C-8 to give primarily a *trans* ring junction. The precise distribution of *trans:cis* epimers was not established, but the minimum distribution of *trans* C:D isomer is estimated at 50%. This is comparable to the yield of *trans* isomer obtained by reduction of the corresponding β-oriented *t*-butoxy group at C-8 [see Z.G. Hajos, R.A. Micheli, D.R. Parrish and E.P. Oliveto, *J. Org. Chem.*, 32, 3008 (1967)] and from reduction of a similar system bearing a β-oriented hydroxyl at C-8 with a propionic acid residue at C-4 [see L. Velluz, G. Nominé, G. Amiard, V. Torelli, and J. Cérede, *Compt. Rend.*, 257, 3086 (1963)].

IIIF-2   3-METHOXY-17β-CARBOXYOESTRA-1,3,5(10),6,8-PENTAENE

D. K. Banerjee, N. Mahishi, and D. Devaprabhakara; *Tetrahedron Letters*, 479 (1968).

TITLE COMPOUND

IIIF-2   The basic stereochemical step comes when the 14-15 double bond is catalytically reduced to give the *trans* C-D junction [cf. W. S. Johnson, J. W. Peterson, and C. D. Gutsche, *J. Am. Chem. Soc.*, 69, 2942 (1947)]. Reduction of a 14-15 double bond, reduction of a hydrindenone containing a large substituent at C-4 (see Synthesis IIIF-1), and ring contraction of a *trans* fused decalin [see R. B. Woodward, F. Sondheimer, D. Taub, K. Heusler, and W. M. McLamore, *J. Am. Chem. Soc.*, 74, 4223 (1952)] constitute the three general approaches to the synthesis of C-D *trans* fused systems.

IIIF-3 RETROPROGESTERONE AND DEHYDRORETROPROGESTERONE

A. K. Krubiner, G. Saucy, and E. P. Oliveto; *J. Org. Chem.*, **33**, 3548 (1968).

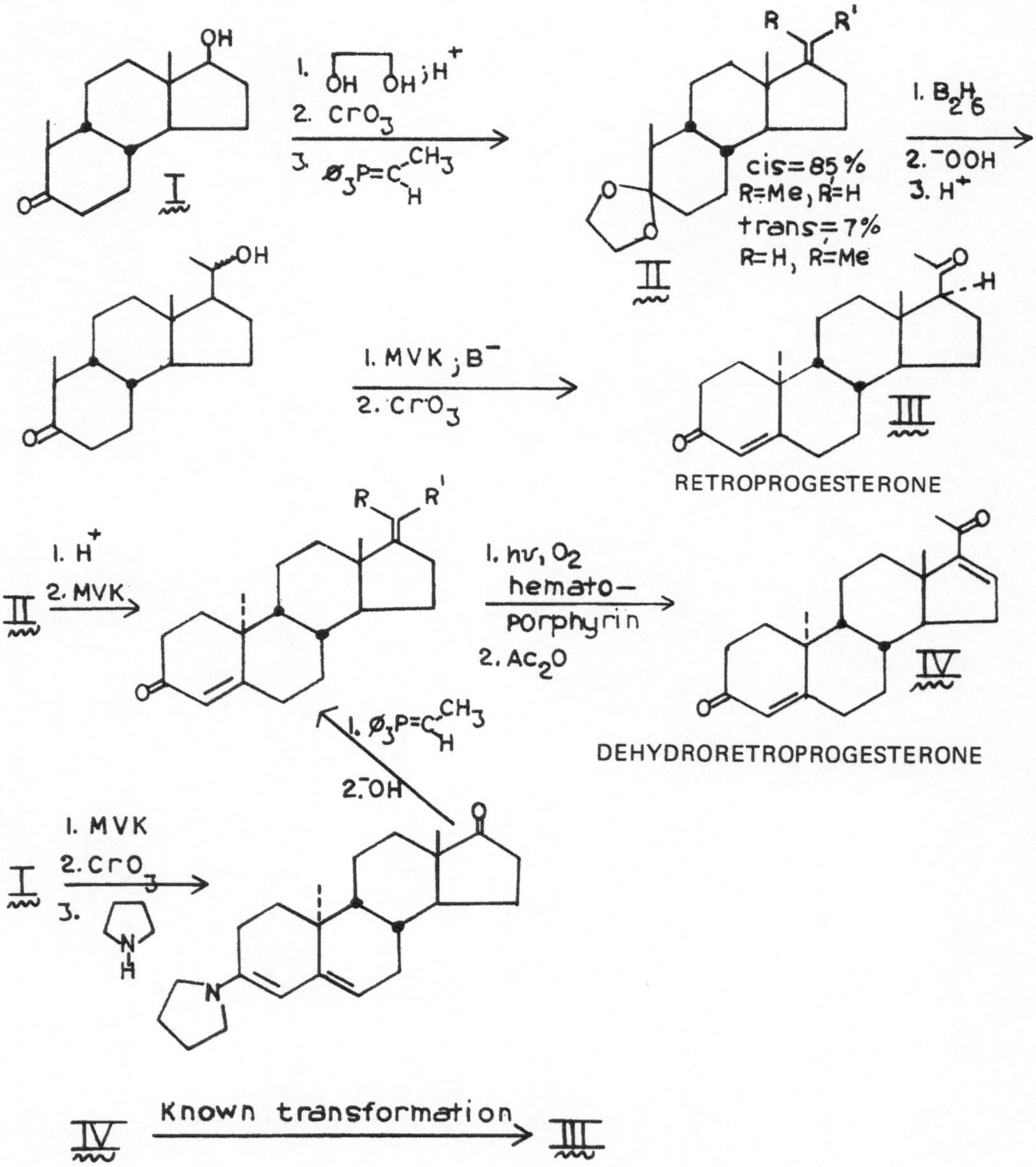

RETROPROGESTERONE

DEHYDRORETROPROGESTERONE

$$\underset{\sim}{IV} \quad \xrightarrow{\text{Known transformation}} \quad \underset{\sim}{III}$$

IIIF-3   Compound **I** arises from catalytic reduction of the final product in Synthesis IIIF-1. Michael addition to methyl vinyl ketone occurs in a convex sense to the *cis* fused decalin [cf. G. I. Poos, G. E. Arth, R. E. Beyler, and L. H. Sarett, *J. Am. Chem. Soc.*, 75, 422 (1953)] leading, in this case, to the abnormal stereochemistry of the C-19 methyl group.

IIIF-4 ESTRONE

C. H. Kuo, D. Taub, and N. L. Wendler; *J. Org. Chem.,* **33**, 3126 (1968).

ESTRONE

IIIF-4   The concept behind this synthesis dates back to Russian work wherein the starting vinyl tetralol was condensed with methyldihydroresorcinol in the presence of Triton B to give a tricyclic diketone which is formally the result of an S_N2' displacement of hydroxide [see S. N. Ananchenko and I. V. Torgov, *Dokl. Akad. Nauk. SSSR*, 127, 553 (1959)]. The authors in this paper present convincing evidence that the condensation is autocatalytic, i.e., the β-diketone serves as the proton donor to aid in the departure of the hydroxyl group. Indeed, no condensation occurs if a 1:1 ratio of base:β-dicarbonyl system is employed. A significant practical advance is the use of a stable isothiuronium salt which, unlike the tetralol, can be readily stored and which condenses with the dione in aqueous ethanol in 90% yield.

IIIF-5 16,17-DEHYDROPROGESTERONE

W. J. Johnson, M. F. Semmelhack, M. U. S. Sultanbawa, and L. A. Dolak; *J. Am. Chem. Soc.*, **90**, 2994 (1968).

16,17-DEHYDROPROGESTERONE

IIIF-5   This synthesis is a splendid culmination of intensive studies in stereospecific polyene cyclizations (cf. fichtelite Synthesis IIC-4) and stereospecific synthesis of trisubstituted olefins (cf. IIIC-2) conducted by the Johnson group. In this synthesis, a symmetrical dimethylcyclopentenol (cf. symmetrical dimethylcyclohexenol in the fichtelite case) is used to generate a cation which triggers the polyene cyclization. The latter reaction proceeds in ca. 30% yield to give a tetracyclic system in which five contiguous asymmetric centers are properly placed. The concurrent ring expansion of the A ring and ring contraction of the D ring provides a fitting climax to this work.

IIIG-1 DIDEOXYZEARALANE

H. I. Wehrmeister and D. E. Robertson; *J. Org. Chem.,* **33**, 4173 (1968).

DIDEOXYZEARALANE

IIIG-1   The key steps in this synthesis are the first and last. The first step is a Perkin condensation of 10-undecenoic anhydride with phthalic anhydride to give a required phthalide. It could be imagined that orientational problems could arise if this condensation were applied to a 2,4-dioxygenated anhydride as required for zearalenone. A Brown-type hydration of the terminal double bond [cf. H. C. Brown and P. Geoghegan, *J. Am. Chem. Soc.*, 89, 1522 (1967)] produces the required secondary alcohol after reductive demercuration ($NaBH_4$).

The fourteen-membered lactone is closed in 24% yield. It will be noted that the closure need not necessarily involve the intermediacy of a hydroxy acid chloride. The active acylating agent could well be the mixed chlorocarbonate anhydride.

IIIG-2 ZEARALENONE

D. Taub, N. N. Girotra, R. D. Hoffsommer, C. H. Kuo, H. L. Slates, S. Weber, and N. L. Wendler; *Tetrahedron*, **24**, 2443 (1968).

ZEARALENONE

IIIG-2   Treatment of 6-methyl-3,4,5,6-tetrahydro-α-pyrone with the Grignard reagent from 5-bromo-1-pentene gives a hydroxyketone which cyclizes to an enol ether. The latter is converted to an acetal which serves to protect the hydroxyl and ketonic groups necessary to build zearalenone.

The Wittig reaction is conducted using the sodium salt of 2,4-dimethoxy-O-phthaldehydic acid. The use of the latter prevents lactonization of the betaine intermediate formed in the Wittig reaction.

After the Wittig closure, the acetal is cleaved on workup. The fourteen-membered lactone is obtained in ca. 10% yield by a mixed anhydride closure.

IIIG-3 ZEARALENONE

L. Vlattas, I. T. Harrison, L. Tökés, J. H. Fried, and A. D. Cross; *J. Org. Chem.*, **33**, 4176 (1968).

ZEARALENONE

IIIG-3   Special steps (saponification of an intermediate acetal) had to be taken to decarbethoxylate the intermediate 6-carbethoxy-5,9-dioxo-1-decene to avoid aldolization. This allowed for decarboxylation under mild conditions (aqueous dioxane-tosyl acid-room temperature).

It is interesting to note that selective deketalization of the C-10 ketal was possible in the benzoic acid systems but not in the ethyl benzoate derivative.

AUTHOR INDEX

Aaron, H.S. 220
Abe, K. 106
Achiwa, K. 235
Afonso, A. 115, 163, 180
Ager, J.H. 10
Agui, H. 14
Amiard, G. 243
Ananchenko, S.N. 249
Andersen, N.H. 234, 236
Anjaneyulu, B. 209
Aoki, T. 37
Arth, G.E. 247
Autrey, R.L. 18
Axen, U. 238
Ayer, W.A. 56

Bader, F.E. 195
Badger, R.A. 153, 161
Bailey, D.M. 83
Balashova, E. 204
Ban, Y. 38, 48
Banerjee, D.K. 244
Bárczai-Beke, M. 16
Barton, D.H.R. 47, 184, 186
Batcho, A.D. 190
Beck, J. 90
Beck, J.R. 2
Bellin, S. 110
Berger, C. 148
Bergess, D. 139
Beyler, R.E. 247
Bhattacharya, S.K. 209
Bianco, E.J. 194
Bickel, H. 195
Biere, H. 20
Birch, A.J. 102
Booher, R.N. 2
Bose, A.K. 208, 209
Bowman, W.R. 56
Bozzato, G. 136
Brady, S.F. 225
Bredenberg, J.B. 115, 163
Britton, R.W. 150
Brossi, A. 70
Brown, A.C. 2
Brown, H.C. 253
Brown, M. 126, 153

Büchi, G. 37, 39, 84, 85, 104, 130, 200
Bundy, G.L. 138
Burckhardt, U. 51
Burgstahler, A.W. 37
Burness, D.M. 131
Butler, K. 194
Bylsma, F. 90

Cable, J. 86
Campbell, S.F. 223
Cardillo, G. 214
Carlson, J.A. 200
Carlson, R.M. 234
Castedo, L. 92
Cavanaugh, R. 173
Cérede, J. 243
Chan, D. 126
Chan, K.K. 88
Chang, D.T.C. 171
Chappell, R.L. 114
Chetty, G.L. 134
Cheung, H. 112
Christman, K.F. 211
Claussen, U. 218
Coates, R.M. 122, 146, 147
Coffen, D.L. 37, 85
Cohen, N. 50
Cohen, T. 7
Conover, L.H. 194
Corey, E.J. 149, 161, 224, 225, 234, 235, 236
Cornforth, J.W. 225
Cornforth, R.H. 225
Coverdale, C.E. 51
Cox, J.M. 51
Cram, D.J. 225
Cretney, W.J. 90
Cricchio, R. 214
Crombie, L. 210, 217
Cross, A.D. 256
Curphey, T.J. 61, 225
Czajkowski, R.C. 190

Dadson, B.A. 12
Danieli, N. 115
Danishefsky, S. 158, 167, 173
Dave, K.G. 26, 52
Daves, G.D. 212

Dawson, D.J. 139
deJongh, H.A.P. 51
Dev, S. 116
Devaprabhakara, D. 244
De Waal, W. 150
Diana, G.D. 30
Djerassi, C. 137
Dolak, L.A. 250
Dolby, L.J. 20
Dolfini, J.E. 13
Donohue, J. 197
Dumas, D. 158
Dunitz, J.D. 197
Dutta, P.C. 172
Dyke, S.F. 74

Eaton, P.E. 153
Eby, C.J. 69
Edwards, J.A. 222
Elhafez, F.A.A. 225
Ellis, M.C. 8
Enggist, P. 104
Erickson, B.W. 224
Erman, W.F. 108
Etheredge, S.J. 240

Failli, A. 88
Fanta, W.I. 108, 138
Faulkner, D.J. 50, 223
Ferguson, C.P. 220
Ficini, J. 151
Flaugh, M.E. 228
Folkers, K. 212
Foster, G.H. 12
Franck-Neumann, M. 148
Frey, A.J. 195
Friary, R.J. 162, 178
Fried, J. 240
Fried, J.H. 256
Friis, P. 212
Fromson, J.M. 88
Fujimoto, Y. 38
Fukumoto, K. 6, 14, 43, 44, 46, 80

Gama, Y. 32
Gandhi, S.S. 78
Ganem, B.E. 225
Gaoni, R. 3
Gaoni, Y. 215
Gaskell, A.J. 22
Geoghegan, P. 253
Gervay, J.E. 45
Giarrusso, F. 176
Gilman, N.W. 224, 225
Ginsberg, D. 41
Girotra, N.N. 254
Gladstone, W.A.F. 86
Gletsos, C. 88
Gnewuch, C.T. 26

Gosh, R. 219
Grafen, P. 30
Graham, D.W. 51
Grethe, G. 70
Gupta, D.N. 152
Gurbaxani, S. 112
Guthrie, R.W. 83
Gutsche, C.D. 83, 245

Habicht, E.R. 50
Hajos, Z.G. 242, 243
Hamon, D.P.G. 50
Hanauer, R. 230
Hanssen, H.W. 86
Harding, K. 236
Hardtmann, G. 196
Harley-Mason, J. 12, 92
Harman, M. 239
Harrison, I.T. 256
Hauser, C.R. 69
Hayasaka, T. 14
Heathcock, C.H. 128, 142, 153, 161
Heckman, R. 110
Hege, B. 110
Heim, S. 240
Hemesley, P. 210
Herz, W. 182
Heusler, K. 187, 195, 245
Hiiragi, M. 14
Himizu, J. 240
Hirai, S. 36, 37
Hirata, Y. 198
Hisamitsu, T. 154
Ho, P. 34
Hochstetler, A.R. 202
Hoffsommer, R.D. 254
Holden, K.G. 239
Hooz, I. 168
Horibe, I. 120, 121
Huffman, R.W. 174
Hwang, B. 239
Hyeon, S.B. 100

Ibuka, T. 40
Ichihara, A. 154, 155
Ichikawa, H. 100
Ilton, M.A. 225
Inubashi, Y. 81
Ireland, R.E. 125, 139, 176, 177
Irie, H. 24, 65
Ishimaru, H. 14
Isoe, S. 100
Ito, N. 154

Jackson, A. 22
Jackson, A.H. 232
Jacobson, A.E. 10
Jautelat, M. 149
Jensen, N.P. 168
Johnson, F. 17
Johnson, P.C. 140

Johnson, W.S. 50, 51, 127, 168, 169, 223, 225, 245, 250, 251
Johnston, J.D. 194
Joseph, T.C. 56
Joule, J.A. 22
Just, G. 239

Kabbe, H.J. 30
Kadin, S.B. 137
Kagawa, S. 154, 155
Kakisawa, H. 166
Kakizawa, M. 32
Kamada, T. 154
Kametani, T. 6, 14, 43, 44, 46, 80
Kane, V.V. 216
Kaneko, C. 115, 163
Kanematsu, K. 10
Kathawala, F. 196
Kato, K. 198
Katsumura, S. 100
Katz, T.J. 183
Katzenellenbogen, J.A. 224, 225
Kawakami, E. 32
Kawata, K. 36, 37
Keeley, S.L. 60
Keeton, R. 102
Kelly, T.R. 128, 142
Kenner, G.W. 232
Kessar, S.V. 78
Keziere, R.J. 124
Kierstead, R.C. 177
Kierstead, R.W. 195
Kigasawa, K. 14
Kim, H.L. 61
Kishimoto, T. 65
Kitano, M. 40
Klem, R. 5
Kobayashi, K. 51
Kocsis, K. 37, 85
Korst, J.J. 194
Korte, F. 218
Kovats, E. 104
Kretchmer, R.A. 55
Kröhnke, F. 77
Kropp, P.J. 141
Krubiner, A.K. 246
Kugita, H. 62
Kulsa, P. 84
Kumar, B. 4
Kumari, D. 186
Kuo, C.H. 248, 254
Kusama, O. 80
Kutney, J.P. 86, 88, 90
Kwok, R. 2

Lahey, F. 42
Leeney, T.J. 92
Leimgruber, W. 190
Leopold, E.J. 168
Lewin, A.H. 7
Li, T. 30, 223
Lier, E.F. 184

Lin, C.H. 240
Lin, M.S. 66
Lincoln, F.H. 238

McCapra, F. 45
McGhie, J.F. 184
McGillivray, G. 232
McLamore, W.M. 77, 187, 195, 245
McLean, S. 66
McMurry, J.E. 160
Maeda, K. 206
Mahajan, R.K. 78
Mahalanabis, K.K. 172
Mahishi, N. 244
Mankas, M.S. 208, 209
Marshall, J.A. 118, 138, 139, 140, 144, 202
Maruishi, T. 166
Masaki, Y. 81
Masamune, T. 51
Mathew, K.K. 225
Matsui, M. 170
Matsumoto, S. 154, 155
Matsumoto, T. 32, 154, 155
Matsunaga, M. 32
Mattingly, G.S. 76
May, E.L. 10
Mazur, Y. 115
Mechoulam, R. 3, 215
Merlini, L. 214
Meyer, W.L. 164, 174, 175
Micheli, R.A. 243
Mikol, G.J. 165
Miles, D.H. 4
Minato, H. 120, 121, 156
Mirrington, R. 182
Mitra, R.B. 161
Money, T. 45
Moobery, J.B. 196
Moon, B.J. 74
Mori, K. 170
Morimoto, A. 96
Mukhopadhyay, S.K. 172
Mummenhoff, P. 218
Murai, A. 51
Murayama, S. 126
Musil, V. 54
Muxfeldt, H. 196, 197

Nabors, J. 4
Nagai, M. 48
Nagasaki, T. 156
Nagata, W. 36, 37, 85
Nandin, I.C. 37
Nelson, V.R. 88
Nishida, S. 154, 155
Noda, K. 96
Nominé, G. 19, 58, 243

Ogiso, A. 171
Ohashi, M. 166, 167
Oishi, T. 38, 48
Oh-ishi, T. 62

Ohno, M. 161, 206
Okumura, T. 36
Okuno, T. 32
Oliveto, E.P. 242, 243, 246
Omi, J. 32
Onaka, T. 97
Ourisson, G. 148

Padeiskaya, E. 204
Pandey, R.C. 116
Pappas, S.P. 5
Parfitt, R.T. 10
Parrish, D.R. 242, 243
Partridge, J.J. 118
Pattenden, G. 210, 226
Patterson, J.W. 153, 161
Patterson, L.E. 2
Paust, J. 234
Pawson, B. 112
Pelletier, S.W. 114, 132, 171
Pershin, G. 204
Pesaro, M. 136
Peterson, J.W. 245
Pfister, G. 125
Philipp, A. 34, 35, 82
Piers, E. 124, 150
Pike, J.E. 238
Pike, M.T. 144
Pohland, A. 2
Ponsford, R. 217
Ponticello, I. 68
Poos, G.I. 247
Posler, J. 162
Posner, G.H. 225
Potts, K.T. 76
Powers, J.C. 68
Prabhakar, S. 114, 132
Pranc, P. 2
Preobrazhenskaya, M. 204

Radlick, P. 5
Rampal, A.L. 78
Rapoport, H. 192, 228
Razdan, R.K. 216
Rizzi, G.P. 50
Roberts, D. 110
Robertson, D.E. 252
Rockey, B. 2
Rodriguez, H.R. 63
Rogalski, W. 197
Roman, S.A. 224
Roos, O. 30
Rosati, R.L. 84
Russell, G.A. 165

Sainsbury, M. 74
Saito, K. 154
Sakan, F. 154, 155
Sakan, T. 100, 106
Sallay, S.I. 161
Santhanakrishnan, T.S. 240

Sarett, L.H. 247
Sasaki, F. 206
Satoh, F. 6, 46
Saucy, G. 112, 246
Schlessinger, R.H. 55
Schlosser, M. 211
Schmiegel, W.W. 178
Schnautz, N. 4
Schneider, W.P. 238
Schroeder, P.G. 174
Schudel, P. 136
Schulenberg, J.W. 115
Schwartz, M.A. 162
Scott, A.I. 45
Scott, J.W. 51
Scullard, P.W. 18
Seebach, D. 235
Semmelhack, M.F. 250
Shamma, M. 63
Sharma, G.M. 45
Shaw, J.E. 122, 146, 147
Shew, D.C. 164
Shin, H. 154, 155
Shioiri, T. 94
Shirahama, H. 154, 155
Siddall, J.B. 222
Simeone, J.F. 178
Simonovitch, C. 239
Sims, J. 5
Slates, H.L. 254
Smith, H. 192
Smith, K.M. 232
Smith, P. 56
Soh, K. 42
Sondheimer, F. 115, 187, 195, 245
Sonnet, P.E. 37, 85
Spencer, T. 178
Spencer, T.A. 162
Spiegelman, G. 208
Sprague, P.W. 26
Spurlock, S. 5
Stevens, R.V. 8, 28, 52, 59
Stork, G. 13, 55, 57, 115, 151, 167
Sugahara, H. 14
Sugahara, T. 43
Suhara, Y. 206
Sultanbawa, M.U.S. 250
Sunder-Plassmann, P. 240
Suvorov, N. 204
Szántay, C.S. 16

Tahara, A. 115, 163
Tahk, F.C. 60
Takasugi, H. 96
Takasugi, M. 51
Taub, D. 187, 195, 245, 248, 254
Thompson, J.L. 238
Todd, A.R. 219
Tökés, L. 256
Tomita, M. 40
Torelli, V. 243

Torgov, I.V. 249
Torupka, E.J. 86
Trueblood, K.N. 197
Turchin, K. 204
Turner, R.B. 30

Uhde, G. 104
Umezawa, H. 206
Uskokovic, M. 70
Uvarova, N. 204
Uyeo, S. 24, 65, 82

Valenta, Z. 82, 83
Valls, J. 19, 58
van Tamelen, E.E. 5, 169, 225
Vatakencherry, P.A. 161
Vedejs, E. 196, 234

Velluz, L. 19, 58, 243
Villarica, R.M. 162
Vlattas, I. 234, 236, 256

Wakamatsu, T. 38
Wall, E.N. 222
Warnock, W.D.C. 86
Watanabe, T. 96
Watt, D.S. 178
Weaver, T.D. 162
Weber, S. 254
Webster, M.S. 197
Wegfahrt, P. 192
Wehrmeister, H.I. 252
Weinstock, J. 239

Weisbach, J.A. 239
Welzel, P. 186
Wendler, N.L. 248, 254
Wenkert, E. 26, 27, 52, 53, 115, 139, 163
Wentland, M.P. 8, 28, 59
Wharton, P.S. 225
Whelan, J. 66
White, J.D. 152
Whitesides, T. 5
Whitlock, H.W. 51, 230
Wiesner, K. 34, 35, 54, 82, 83
Wiesner, K.J. 54
Wilds, A.L. 137
Williams, K.R. 239
Wilczynski, J.J. 212
Wilson, N.D. 22
Winter, R.E.K. 234
Woodward, R.B. 77, 183, 187, 194, 195, 229
Wright, D.C. 219
Wüest, H. 130

Yagi, H. 6, 14, 46
Yagnitinsky, B. 215
Yamada, S. 94
Yanagiya, M. 32
Yasuda, S. 32
Yoshitake, A. 24
Young, H. 182
Zalkow, L.H. 4, 134
Zalkow, V.B. 134
Ziegler, F.E. 37, 72, 85
Zoretic, P.A. 72
Zurfluh, R. 222

COMPOUND INDEX

Abscisic acid 110
17-Acetyl-5α-etiojerva-12,14,16-trien-3β-ol 50
(+)-Acetylideneisodihydroiresin 132
(−)-Acetylideneisioresin 132
Acronycine 2
Actinidiolide 100
Actinidol 100
α-Agarofuran 144
β-Agarofuran 144
Ajaconine, intermediate 4
Alloevodionol 214
β-Amyrin 184
Androcymbine 6
Androcymbine type compounds, 6
Andrographolide Lactone 114
Anthramycin 190
Anthramycin methyl ether 190
Apoferrorosamine 8
Ar-himachalene 116
$\Delta^{1,10}$-Aristolene 146
Aristolone 148
Aspidosperma alkaloid model 26
Atidine intermediate 4

6,7-Benzomorphan 10
α-Bourbonene 152, 153
β-Bourbonene 152
Bulnesol 118

Cannabichromene 214, 216
Cannabicyclol 216
Cannabigerol 214, 216
Carabrone 120
Carnosic acid dimethyl ether 164
Carnosol dimethyl ether 164
Cinerolone 210
Condyfoline 12
Coreximine 14
(−)-Corynantheidine 16
(3S, 15S, 20R)-Corynantheine 18
10-Cyano-12-hydroxy-7-oxo-17-norpodocarpa-5,
   8,11,13-tetraene 174
Cycloartenol 186
Cyclopenin 192

Dasycarpidone 20, 22

Dehydroelsholtzione 104
Dehydrofuropelargones 130
16,17-Dehydroprogesterone 250
Dehydroretroprogesterone 246
4-Demethylaristolone 150
6-Demethyl-6-deoxytetracycline 194
Deoxophylloerythroetioporphyrin 228
Desoxypodocarpic acid 176
Dideoxyzearalane 252
Dihydroactinidiolide 100
Dihydrocrinine 24
Dihydrogambirtannine 26
Dihydroiresin 132
Dihydroradicinin 198
Dihydrovittatine 24
15,16-Dimethoxyerythrinan-3-one 28
Diterpene alkaloids, approaches to 30, 31, 32, 34

Epidasycarpidone 20, 22
Epiibogamine 36, 37, 38
Epilupinine 52
12-Epilycopodine 54
Epipatchouli alcohol 158
11-Epiprostaglandin E_1 236
11,15-Epiprostaglandin E_1 236
15-Epiprostaglandin E_1 236
Epivincadine 88
Eremoligenol 122
Eremophil-3,11-diene 124
Eremophilene 122
trans, cis-6-Ethyl-10-methyldodeca-5,9,-dien-2-
   one 222
Estrone 248
β-Eudesmol 128, 134

Ferruginol 166
Fichtelite 168
Franklinone 214
Fulvoplumierin 200

Geosmin 202

Hasubanan alkaloids, ring systems related to 40
Hernadine 42
Homoerythrinadienone 44
Homoproaporphine type compounds 46
Hydrojulolidine ring system 53

264

Hydrolulolidine derivative 52
(−)-17β-Hydroxy-Δ$^{9(10)}$des A-androsten-5-one 242

Ibogamine 36, 37
Illudane, functionalized 154
Illudin M 155
Indolmycin 204
Isodihydroiresin 132
Isoiresin 132
(−)-Isoiresin diacetate 132

Jasmolone 210
Juniper Camphor 134
Juvabione 112
Juvenile hormone, cecropia 223, 224

Kasugamycin 206
Kaur-16-en-19-oic acid 170

Lindestrene 156
Lupinine 52
Lycopodine 55, 56

Mesembrine 59, 60, 61, 62, 63
3-Methoxy-17β-carboxyestra-1,3,5(10),6,8-penta-ene 244
13-Methoxypodocarpic acid 176
Methyl Deisopropyldehydroabietate 162
1-Methyl-16-demethoxycarbonyl-20-desethyli-dene-vobasine 94
Methyl-2-hydroxy-1,1,4a-trimethyl-7-oxoperhyd-rophenanthrene-8a-carboxylate 172
Methyl podocarpate 178
Methyl trachylobanate 182
Minovine 88
Myosmine 8

Nezukone 102
Nootkatone 136
Norbourbonone 152
Nor-ketoagarofuran 142

Octaethylporphyrin 230
Ochotensimine 65, 66
Ochotensine 65
7-Oxaprostaglandin F$_1$α 240
Oxyporphyrins 232

Patchouli alcohol 158
5,6-*trans*-Penicillin V methyl ester 208
Perhydrojulolidine 53
Perhydrojulolidone 53
Perlolidine 68
Petaline 70

Pluviine 52
Podocarpic acid 178
Propylure 226
Prostaglandin A$_1$ 234
Prostaglandin E$_1$ 234, 236, 238
Prostaglandin F$_{1\alpha}$ 234
Prostaglandin F$_{1\beta}$ 234
Pyrethrolone 210

Quebrachamine 88
Quebrachamine skeleton 72

Retroprogesterone 246
Rhodoquinone 212
Rosefuran 104

Sabina ketone 108
Sabinene 108
Sabinene hydrate 108
Sandaracopimaric acid 180
Sanguinarine 74
Sativene 160
Selin-11-en-4α-ol 134
Sempervirine hydrobromide 76
Solanidine 78
Stebisimine 80
Strychnos alkaloid model 26, 48

Terramycin 196
Tetrahydrocannabinol congener 220
Δ6a,10a-Tetrahydrocannabinol 218
Tetrahydroeremophilone 126
Tubifolidine 12
Tubifoline 12

Valeranone 138
Veatchine 82
Velbanamine 84
Verarine 86
Veratramine 50 Verbenalol 106
β-Vetivone 140
Vinca alkaloids, dimeric 90
Vinca alkaloids, monomeric 88
Vincadifformine 88
Vincadine 88
Vincaminol 92
Vobasine skeleton 94
Vincaminoreine 88
Vincaminorine 88
Vindoline 90

Withasomnine 96, 97

Zearalenone 254, 256